# Focus on GIS Component Software

**Featuring ESRI's MapObjects®**

*Robert Hartman*

**Focus on GIS Component Software:**
**Featuring ESRI's MapObjects®**

Robert Hartman

Published by:
OnWord Press
2530 Camino Entrada
Santa Fe, NM 87505-4835
USA

First Edition, 1997
SAN 694-0269

10 9 8 7 6 5 4 3 2 1

Printed in the United States of America


**Library of Congress Cataloging-in-Publication Data**

Hartman, Robert, 1962-
Focus on GIS component software : featuring ESRI's MapObjects / Robert Hartman.
p.    cm.
Includes index.
ISBN 1-56690-136-7
1. Geographic information systems. I. Title.
G70.212.H37 1997
910'.285—dc21        97-20769
                          CIP

## Trademarks

MapObjects is a registered trademark of Environmental Systems Research Institute. OnWord Press is a registered trademark of High Mountain Press, Inc. Many other products and services mentioned in this book are either trademarks or registered trademarks of their respective companies. OnWord Press and the author make no claim to these marks.

## Warning and Disclaimer

This book is designed to provide general information about geographic information systems. Every effort has been made to make this book complete, and as accurate as possible; however, no warranty or fitness is implied.

The information is provided on an "as-is" basis. The author and OnWord Press shall have neither liability nor responsibility to any person or entity with respect to any loss or damages in connection with or arising from the information contained in this book.

## About the Author

Robert Hartman is a principal consultant with the San Diego Processing Corporation, a Southern California-based technology consulting firm. He has a B.A. in geography from California State University at Fullerton, and an M.B.A. with an emphasis in strategic information systems from Heriot-Watt University in Edinburgh, Scotland. Since 1985 he has provided services ranging from programming to project management for clients seeking to integrate GIS technology with other information technologies in the development of strategic business systems.

## Acknowledgments

This book is the result of the direct and indirect contributions of many people, only a few of whom can be mentioned here. Thanks to those who contributed time to review the manuscript and discuss ideas, including Kent Wright, Technical Manager of GIS at San Diego Data Processing Corporation (SDDPC); Bruce Joffe of GIS Consultants; Bill Bamberger, Marketing Manager of GIS at SDDPC; and Russell Louks of Environmental Systems Research Institute (ESRI).

Thanks to Mark Sorensen, President of Geographic Planning Collaborative, for hours of productive conversation regarding the meaningful use of technologies and concepts discussed in this book. Likewise, thanks to Ken Blue, President of Blue Inc. and Senior Consultant at San Diego Consulting Group, for his insights on a wide variety of issues.

Thanks to those who have participated with me in testing the ideas in this book against the real world: Jody Betry, Dale Brooks, Krista Dill, Ron Harris, Bill Lull, Brian McKee, and many others. Thanks to Lisa Stapleton and the San Diego Geographic Information Source (SanGIS, formerly RUIS) for support. Also thanks to Glenn O'Grady and Huan Nguyen for contributing to this work by providing guidance so many years ago.

Thanks to the staff at ESRI, especially Bernie Szukalski for his support of this project, Russell Louks for enlightening me on some important technical issues, and Karen Hurlbut for her investigative assistance. Thanks to Kirk Fisher of Oracle for putting me on the trail of Oracle's NCA and spatial components. Many thanks to the staff at High Mountain Press, including Barbara Kohl and David Talbott for their openness to new ideas and approaches, and especially to Daril Bentley for being a patient and helpful editor far above and beyond the call of duty.

A special thanks to my wife for her patience and support in this and a million other endeavors.

**OnWord Press**

OnWord Press is dedicated to the fine art of professional documentation. In addition to the author, who developed the material for this book, other members of the OnWord Press team helped make the book end up in your hands.

Dan Raker, President and Publisher
David Talbott, Vice President, Development and Acquisitions
Dale Bennie, Vice President, Publishing
Rena Rully, Publisher
Carol Leyba, Associate Publisher
Scott Brassart, Acquisitions Editor
Barbara Kohl, Associate Editor
Daril Bentley, Senior Editor
Cynthia Welch, Production Manager
Michelle Mann, Production Editor
Lynne Egensteiner, Cover Designer, Illustrator

*To Hillary, Riley, and Chase*

# Contents

## Chapter 6: Inside a Geographic Software Component: A MapObjects Overview   119

## Chapter 7: Creating Maps: MapObjects Maps and Layers   155

## Chapter 8: Displaying Maps:
## MapObjects Symbols and Renderers     **193**

## Chapter 9: Analyzing Maps: MapObjects, Recordsets, Geometric Objects, and Spatial Queries      237

## Chapter 10: Looking Ahead at Software Components    293

## Appendix A: Other MapObjects Products    315

## Appendix B: For Further Reference    323

## Index    331

# Introduction

## Purpose of the Book

This book is an introduction to the subject of component software, with a particular focus on its impact and use in the field of geographic information systems (GIS). Software components, often referred to as distributed objects or componentware, are essentially "plug-and-play" software objects that easily combine and communicate with one another. More specifically, they are developed according to emerging component standards that enable them to work together across networks, languages, development tools, hardware, and operating systems.

Combined with modern visual programming environments, such as Visual Basic and Delphi, which provide frameworks within which software components can be gathered, components are creating a new approach to software development. This new approach emphasizes the assembly of software applications instead of their construction.

In contrast to the typical approach of developing applications from scratch, a component approach allows developers to acquire pre-built software components and assemble them into new applications. Components have been described as the software counterpart to integrated circuit (IC) hardware. Just as hardware configurations can be built and updated by plugging in ICs, component-based applications can be modified with relative ease by exchanging existing components in the application with newly acquired ones.

This book explores how GIS technology is incorporating the concepts of software components to provide new products and new possibilities for the application of GIS. GIS software components service the increasing demand for GIS capabilities in applications and contexts that are not GIS specific. Such applications may range from the delivery of GIS via the Internet to the inclusion of GIS in task-specific mobile computers and personal digital assistants. Integrating GIS into such situations was difficult, if not impossible, prior to the advent of GIS software components.

This trend of GIS integration through the use of software components is placing new demands on GIS professionals. At the very least, GIS professionals are challenged to be aware of this new approach and what it can provide. Those facing the task of incorporating such products into their tool sets and developing solutions with them are confronted with a wide variety of foreign concepts and terms, such as *object request broker* (ORB), *ActiveX, COM, OLE, CORBA*, and *IIOP*.

This book is intended to assist you in understanding what software components are, why they are important to GIS and to any business that relies on geographic data, and how GIS software components can be used. Along the way, you will become familiar with the concepts and terminology of software components and their application to GIS. You will also have the opportunity to work with an actual GIS software component contained on the accompanying CD-ROM. After reading this book and working with the tools on the accompanying CD-ROM, you will be prepared to begin exploring the use of GIS in applications and situations you may have never imagined possible.

## Audience

The audience for which this book is written is intentionally broad, and consists of decision makers, GIS professionals, and solution developers/programmers. Each of these groups could warrant a separate book specific to their particular interests in the subject of software components. However, component software is so significant a technology to the GIS industry, and there are so few resources currently available regarding it, that this book is structured to be helpful to each of these groups at an introductory level appropriate for each.

The structure of the book allows readers from any of the previously mentioned groups to easily access the material relevant to them, and to extend their reading into as much of the remaining material as desired. Having said this, there are good reasons for decision makers and technical professionals to understand the material in this book that might generally be associated solely with either role.

Managers who fail to understand the technology and technical issues faced by their organizations do so at their own risk. Just as an effective manager needs to possess some knowledge of the mechanics of finance, accounting, and similar management-related disciplines, some knowledge of the mechanics of important technologies is also required. This is particularly true in the current business environment, in which technology processes that lifeblood of today's organization: information. For these reasons, managers and decision makers are encouraged to read the more technical material contained in this book.

On the other hand, technical individuals are being expected to "know the business" to an ever greater degree. Someone once told me that the man who knows "how" will always have a job...working for the man who knows "why." In light of this statement, programmers and other assorted technically oriented professionals are encouraged to avoid underestimating the importance of reviewing the more conceptual and business-oriented material in this book. Understanding the business drivers and impacts of a component approach to software are key to a technical professional's successful implementation of it.

## Structure and Content

As just indicated, this book is intended for decision makers, GIS professionals, and programmers. The book is structured to present the material in increasing level of technical difficulty. The earlier chapters provide more conceptual material. The book then proceeds into technical fundamentals, and closes with several chapters of actual programming examples. Throughout the book, concept summaries are provided in the margins to summarize the content of adjacent paragraphs.

## Book Structure

This structure provides two advantages. First, readers from the three roles addressed as the audience will find in a single location the material most relevant to their particular roles. Managers and decision makers are likely to be most interested in the first portion of the book, technical managers and GIS professionals in the middle and latter portions, and programmers in the final chapters. The exception to this is the final chapter, Chapter 10, which is intended to provide all readers with some perspective on how trends in GIS software components will likely affect them in the future.

Second, the concept summaries provided in the margins throughout the book highlight the important concepts being discussed. By browsing the summaries for any given chapter, you can gain a high-level understanding of the chapter's material without reading the chapter in detail. This "quick review" effect is intended to make the technical material more accessible to decision makers, and to make the business-oriented material more accessible to technicians.

## Chapter Content

This book begins, in Chapter 1, with a review of the business case for software components. The business demands on technology in today's environment are examined, and the attempt of object-oriented technology to respond to these demands is discussed. Software components are introduced as a means of advancing the software industry into an approach toward software development appropriate to today's business environment. In addition, a few brief examples are provided of businesses successfully using software components.

Chapter 2 is a focused discussion on the case for software components in the GIS industry. The need for and benefits of software components in the GIS industry are discussed in the context of an evolutionary view of GIS and its technology as responding to an ever-expanding and increasingly diverse user base. Discussion includes the movement of GIS into the technical infrastructure of organizations and how GIS software components meet these demands.

Beginning with Chapter 3, the discussion shifts from "why" GIS software components are important to "what" GIS software components are. Chapter 3 reviews the differences between typical software objects and software components. The discussion focuses on two critical areas in which the traditional object model is being extended in order to support software components: object interfaces and inter-object communication infrastructure.

Chapter 4 provides an overview of two important software component architectures in today's market. Microsoft's

COM architecture is discussed in some detail to examine how it provides a foundation for creating software components. OLE, DCOM, and ActiveX and other COM-related concepts are described. This chapter also provides an overview of CORBA and how it provides an architecture for component creation and use.

Chapter 5 completes the section on what software components are by introducing GIS software components. The importance of industry-level standards that supplement technology standards (such as DCOM and CORBA) is discussed. The work of the Open GIS Consortium and its OGIS specification are discussed as they relate to such component standards for the GIS industry. Various approaches to current implementations of GIS software components are also mentioned.

Chapter 6 begins a section on "how" GIS software components work. Using ESRI's MapObjects product and Microsoft's Visual Basic development tool, Chapter 6 provides an inside look at an actual GIS software component (MapObjects) based on Microsoft's ActiveX architecture. This chapter discusses the design of MapObjects and how objects in the component are accessed. An evaluation copy of MapObjects is provided on the accompanying CD-ROM.

Chapter 7 provides an example of how to create maps with the MapObjects component. The map creation capabilities of MapObjects are reviewed and you are guided through an actual Visual Basic application, contained on the accompanying CD-ROM, to demonstrate these capabilities.

Chapter 8 reviews MapObjects map display capabilities in a format similar to Chapter 7. You are guided through another sample application, contained on the accompanying CD-ROM, to demonstrate symbolization using symbol and map rendering objects from the MapObjects component.

Chapter 9 completes the discussion of MapObjects by reviewing its map analysis capabilities. This chapter includes another application walk-through, also contained on the CD-ROM, to demonstrate the use of geometric shapes, tables, and records in performing spatial queries.

Chapter 10 concludes the book by examining the future of geographic software components. Important areas of emphasis for GIS components are examined, including the Internet, spatial data servers, and information appliances. A concluding section reviews how GIS software components are affecting various organizational roles, including business managers, technical managers, GIS professionals, and software developers.

## Contents of the CD-ROM

The CD-ROM that accompanies this book contains an evaluation copy of MapObjects, sample MapObjects applications, MapObjects documentation, and Visual Basic 5.0 updates. These are explained in the following sections.

### Evaluation Copy of ESRI's MapObjects

The accompanying CD-ROM includes an evaluation copy of ESRI's MapObjects GIS software component. The evaluation copy is a fully functioning copy of MapObjects that will expire 90 days after installation on your computer. No technical support is available from ESRI for this evaluation copy.

### Sample MapObjects Applications

Also included on the CD-ROM are sample applications that demonstrate the capabilities of MapObjects. Note that these sample applications will be installed only if you indicate during the installation process that you wish to install them. There are sample applications written in Visual Basic, Delphi, PowerBuilder, and Visual C++ . Also provided with the samples are Shapefile data and bitmaps used by the applications.

The projects discussed in Chapters 7 through 9 are found among the Visual Basic sample applications contained on the CD-ROM. A help topic describing each Visual Basic example is available from the MapObjects help system. To view this topic, start MapObjects help, and from the Contents tab select About the MapObjects Samples in the Using MapObjects book.

## MapObjects Documentation

The accompanying CD-ROM also contains several helpful documents regarding MapObjects. Diagrams of the MapObjects object model are included. The object diagrams are available only if you install the samples when installing MapObjects on your computer. Three versions of the object diagram are provided: *mo11diag.doc* (a Microsoft Word document), *mo11diag.vsd* (a Visio drawing), and *mo11diag.pdf* (an Adobe Acrobat file). See the *readmemo.wri* file on the CD-ROM for helpful information on printing these diagrams.

The complete on-line help system of MapObjects is included with the evaluation copy. If you are using Visual Basic 5.0, see the notes for Visual Basic users that follow.

Also included are the complete on-line help system of MapObjects and a document titled *Getting Started with MapObjects,* which provides a brief tutorial. There are versions of *Getting Started with MapObjects* for PowerBuilder 5.0 and Delphi 2.0. See the *readmemo.wri* file on the CD-

ROM for details on where to find this document after installation.

↦ **NOTE:** *Further documentation and current information regarding MapObjects is available at the MapObjects Developer's Connection on the ESRI Web site at:*

**http://www.esri.com/base/products/mapobjects/developerconnection/modeveloper**

ext.htm

## Updates for Visual Basic 5.0 Users

Visual Basic 5.0 differs from previous versions in several important ways that affect the sample applications and the MapObjects help files. If you are using Visual Basic 5.0, you will need to run the *movb5samples.exe* and *movb5help.exe* files located in the vb5update folder on the CD-ROM. These files should be run *after* installing MapObjects.

*Movb5samples.exe* updates the Visual Basic samples in your MapObjects installation directory to be compatible with Visual Basic 5.0. Note that after you update the sample applications by running *movb5samples.exe*, the projects can no longer be loaded into Visual Basic 4.0.

*Movb5help.exe* updates the MapObjects help files with examples specific to Visual Basic 5.0. Running this installation file will also make the MapObjects help files compatible with changes made to the Visual Basic help system in 5.0. If you are using Visual Basic 5.0 and do not run this installation file, you may experience problems using the MapObjects help files in Visual Basic. See the *vb5update.txt* file in the vb5update folder on the CD-ROM for more information.

## Installing Software from the CD-ROM

To install software from the CD-ROM, perform the following steps:

1. Run the *setup.exe* file on the CD-ROM.

2. Follow the directions of the setup program to proceed to the MapObjects Setup Options screen.

3. At the MapObjects Setup Options screen you will be allowed to select the options Typical, Compact, or Custom. If you want to install the sample applications or any optional MapObjects capabilities—such as ARC/INFO coverage support, SDE support, or ESRI cartographic fonts—select the Custom option and click the Next button.

4. If you select the Custom option, you will be presented with the Custom Options Selection screen. Select the custom options you wish to install and click the Next button.

5. The installation program will continue and complete the installation of MapObjects and any custom options you selected.

6. Review the *readme* file that was copied to your MapObjects installation directory during the installation process. Note that MapObjects is installed in C:\Program Files\ESRI\MapObjects unless you specify otherwise during the installation.

7. If you are a Visual Basic 5.0 user, run the *movb5samples.exe* and *movb5help.exe* installation programs located in the vb5update folder on the CD-ROM. These will update the MapObjects help files and sample applications for use in Visual Basic 5.0.

# 1
# The Business Case for Components

The Gartner Group estimates that components will create a $7 billion industry by 2001.

A new approach to software development is arriving. The capabilities and opportunities of this new approach have prompted the Gartner Group [a leading research and consulting firm in the information technology (IT) industry] to estimate that by the year 2001, the trend will create a $7 billion industry. It is "the wave of the future" according to Mike Blechar, Gartner's director of application development research. What is this new approach? It is the use of software components, also termed componentware or distributed objects.

The component approach to software development partitions applications into manageable pieces that can work together despite elements of the computing environment that traditionally limit software interaction, such as differences in hardware platforms, operating systems, and vendor implementations of the software. This independence from traditional computing environment constraints is where components diverge from typical object-oriented technology. (Subsequent chapters cover this concept in more detail.)

Components are software "pieces" that can be assembled into applications.

This independence also makes possible a new "plug-and-play" approach to software development. In such an approach, software components—each providing a required bit of application functionality such as charting, reporting, or database query—can be acquired and "plugged in together" to assemble working applications.

In contrast, applications today are typically developed from scratch or with limited reuse. The ability to assemble new applications and easily modify existing applications by mixing and matching software components opens up new opportunities for the software development industry and those it serves.

Business forces drive software toward a component approach.

Why is this new approach being developed? What are the driving forces behind this trend toward software components? Ultimately, the computing industry responds to the needs of computer users. As more businesses and organizations come to rely on software as a critical element in the production and delivery of goods and services, the computing industry must work harder to fulfill the demands of such users.

Software components are one element of the computing industry's response to these demands. This chapter explores the business forces that are creating a general movement in IT toward the development and use of software components.

## *What Business Demands from Technology*

Today's business environment is experiencing rapid and dramatic change. Conditions that have typically undergone slow and measured change now seem to transform overnight. Customer requirements, competitive conditions, regulatory requirements, political and economic conditions, and many other dimensions of the business environment now change so quickly that it is all an organization can do to keep up, let alone leverage the changes

The business environment is changing dramatically.

for strategic advantage. This situation confronts any organization, whether private, public, or nonprofit.

These changes not only occur at dramatic speed, but are nearly unmanageable in scope. Change is so fundamental and varied that organizations inevitably find themselves attempting to cope with many conflicting demands. As Charles Handy points out in his book *The Age of Paradox*, successful organizations live in a paradox.

The variety and pace of change in the business environment is nearly unmanageable.

"They must be planned," says Handy, "yet flexible, differentiated yet integrated, mass marketers, while at the same time catering to new niches, producing quality but at low cost. They have to reconcile what used to be opposites, instead of choosing between them." As is explained in the material that follows, this has significant implications for the IT that must support organizations in such an environment.

## Speed, Quality, and Flexibility

Speed, quality, and flexibility are fundamental to managing change.

As an example of opposites that must be reconciled, consider the concurrent demands on today's organizations for speed, quality, and flexibility. These three elements are becoming fundamental to the operation of most organizations. As organizations adopt customer service as a central principle of their business strategy, an increasing emphasis is being placed on product and service *quality* and *consistency*.

At the same time, organizations are forced to constantly monitor and adapt to market events by intense competition and the instantaneous broadcast and assimilation of market information. This means that *speed* becomes essential to the development of products and services that meet the demands of a rapidly changing market.

In addition to quality and consistency, a high degree of *flexibility* is required to integrate products and services in a business environment dominated by partnerships, alliances, and mergers. It is no small challenge to simultaneously develop products and services of consistently high *quality*, develop them *rapidly*, and design them for maximum *flexibility*—as represented in the following illustration.

*Today's business environment requires integrating conflicting demands.*

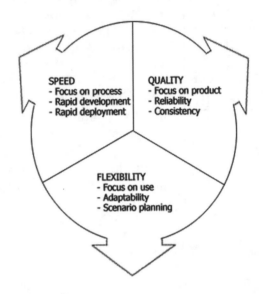

Witness the concurrent demands for speed, quality, and flexibility in the banking industry. Intense competition in an increasingly deregulated market, combined with the nearly real-time flow of news and economic information, requires extremely rapid market response. New financial services are rapidly developed and presented to customers on what seems a daily basis. Many service offerings may have a life span of only several months before being superseded by new offerings—a result of the rapidly changing financial and competitive conditions of the market.

<u>The demands of a change-oriented business environment are often in conflict.</u>

The execution and management of these offerings must be of the highest quality because they deal with an issue very sensitive to most consumers: their money. Mergers of large institutions, such as the recent merger of Chase Manhattan and Chemical Bank, complicate the day-to-day operations of a business even more, because they require large-scale integration efforts to combine complex infrastructures and product offerings. The result is a seemingly impossible challenge to reconcile diverse and apparently opposing requirements.

*Organizations are changing their infrastructures in response to concurrent and conflicting demands.*

Organizations are responding to the dilemma of concurrent and often conflicting demands for speed, quality, and flexibility by changing their infrastructures, which are defined in this context as the people, property, and processes that form the foundation of an organization. Infrastructures that have for many years been oriented toward a more stable business environment are now being modified to match today's change-oriented environment.

For example, organizational structures built to service the relatively static environment of the past are being exchanged for new structures that support quick decision making and easy modification of business processes. Business models are being reinvented to support much shorter product life cycles and more personalized services. Now, in addition to the traditional elements of organizational infrastructure, there is a new element to consider: IT.

## Information Techology in the Infrastructure

*IT is becoming part of the organizational infrastructure.*

In the last few decades, as information has become the lifeblood of organizations, information systems have gradually become embedded at the infrastructure level of many organizations. As an information systems director said in a recent technology advertisement: "I know I've done my job well when information flows as freely and reliably as the electricity in our buildings, and when the network is as invisible as the heating system."

IT is now as fundamental an element in many organizations as the organizational structure or business model. As a result, IT is now required to adapt, along with the rest of the organization's infrastructure, to today's business demands for speed, quality, and flexibility.

There are many elements in an organization's technology infrastructure, including computer hardware, applications, telecommunications, and personnel. The demands for speed, quality, and flexibility manifest themselves in different ways for these elements. For example, personnel must constantly and quickly learn new technologies, and telecommunications networks must support rapid rerouting to accommodate changes in location.

In the area of applications, specifically software development, business demands appear in the form of extremely short software development cycles (speed), high software reliability (quality), and integration with other software and systems (flexibility). The effect of these demands is represented in the following illustration.

| Environment | Market Change | Customer Focus | Mergers and Partnerships |
|---|---|---|---|
| Organization | Speed | Quality | Flexibility |
| Infrastructure | Rapid Development | Reliability | Integration |

Organizations require rapid software development to service rapidly evolving demands. Sometimes the life span of the software developed is as short as, or even shorter than, the development cycle itself. Internal processes or external business services may be so transitory that the software supporting them must be able to be abandoned with minimal cost to the organization.

This concept, sometimes referred to as "throw-away software," recognizes that at times software is a short-term

*Software development is experiencing new demands for speed, quality, and flexibility.*

*The business environment places demands on the organization, which places demands on the IT infrastructure.*

*"Speed" means rapidly developed, low-cost software.*

investment providing immediate gains that contribute to the larger organizational strategy. The Gartner Group has named these "tactical systems." This is a perspective completely foreign to the traditional software development view of all applications as long-term capital investments.

An organization's ability to provide quality products and services is increasingly dependent on its information systems. The software the organization uses must reliably and capably support those services. For example, because most modern banking is conducted electronically, the failure of software supporting large-scale international transactions can affect national and international economies.

*"Quality" means reliable software.*

As a more modest, but still important, example, a "tactical system" supporting a new loan service to small businesses must function to the complete satisfaction of the customer. If it fails to do so, the small businesses will go elsewhere for their loans, and market share is lost.

Applications must be flexible in order to support both integration with other systems and quick response to changing customer requirements. A given software product may need to be integrated with systems inside or outside the organization.

*"Flexibility" means easily integratable software.*

For example, consider the increasingly common practice of reworking an organization's internal support systems, such as accounting, purchasing, finance, and human resources. The integration of these systems with one another is critical to the success of the organization. Any redevelopment or modification of one of these systems must integrate with the other systems. If it does not, the organization can become crippled in its daily operations.

Software integration is also important external to an organization. The systems with which a given software product must integrate may be known or they may be unknown, as in the case of an unforeseen merger. To achieve such a high level of integration, software must be easily modified to adapt to the environment with which it must integrate. Also, rapidly changing customer requirements often demand quick response in order to retain market share. Such quick response necessitates software that can be easily modified to support new requirements.

## The Software Crisis and Object Technology

*Today's software developer has inherited a crisis.*

The new business environment and its consequent demands on IT come at a time when information systems professionals are still struggling to meet the demands of the rapidly disappearing traditional business environment. The limited success of the software development industry in servicing the needs of the traditional business environment has created a problem known as the "software crisis." A brief review of the history of this struggle and attempted solutions will set the stage for examining how software components are providing solutions for the current environment.

## The Software Crisis

*The software crisis: backlogs and defects.*

The "software crisis" refers to huge backlogs of application development projects, combined with significant compromises in the quality of existing software products. In 1991, the software metrics expert Capers Jones declared that the average development project is one year late and 100 percent over budget. A more recent study, conducted in 1994 by the Standish Group International, determined that nearly one-third of corporate information systems projects are canceled during development. Fifty-two percent are completed late, over budget, or with reduced scope.

This leaves a dismal 16 percent of projects successful according to plan. While there are many reasons for such a poor performance on the part of IT, the roots of this crisis are found in a disparity between the methods and practices of the software development industry and those of the business environment it serves.

| 32% | 52% | 16% |
|---|---|---|
| Canceled | Late<br>Overbudget<br>Reduced Scope | Success |

## Software Development as a Craft

Traditional software
development is a craft.

Traditional methods for developing software have been inadequate to service the demands of even the most stable business environments. David Taylor, in his book *Object-Oriented Technology: A Manager's Guide,* makes an excellent case that this is primarily because of an orientation in the software industry toward the software development activity as a craft.

Although nobody would argue with the benefits of attention to detail and individuality that are the hallmarks of a craftsperson, efficiency must also be considered, says Taylor. More specifically, can a craft-oriented approach to software development adequately serve a business environment that long ago traded the craftsman for the assembly-line worker? As Steve Mills, General Manager of Software for IBM has said, software development as it is predominantly practiced today is a medieval art form.

A software craftsperson
builds software from the
ground up.

In the craft-oriented approach, software is largely built "from the ground up," with very little significant reuse or sharing of software developed in earlier efforts. This method is reminiscent of furniture building in preindustrial days. Furniture was built by craftsmen who cut the trees, milled the wood to make lumber, carved the necessary pieces and their decorative detail, fashioned each joint, and assembled the furniture by hand. Because furniture was essentially custom made, the total output of furniture was low, and the quality variable.

It should be obvious that a craft-oriented approach can never adequately service the speed-, quality-, and flexibility-oriented environment previously described. In fact, such an approach cannot even service the traditional business environment and its focus on efficiency. A craft-oriented approach to software development, just like a craft approach to furniture building, results in a low total output of product with variable quality. A craft-oriented approach to software development cannot meet the demands of even the most stable of business environments.

A craft-oriented approach yields limited output and variable quality.

## From Craft to Construction Using Object Technology

At the time software development entered the business environment, that environment had long since moved beyond craft and into construction. Whereas a craft approach focuses solely on the product, a construction approach includes a focus on process, particularly on efficiency. Automobile plants, for example, instituted assembly lines long before the introduction of modern computer hardware and software. Therefore, from the start, a disparity existed between the methods and practices of software development and the environment it served.

Crafted software does not "fit" traditional business's construction approach.

Despite the need for a better match between software development methods and the business environment, the software industry has to a large extent persisted in retaining a craft-oriented approach to its product development. This mismatch may explain the existence of many of the problems in recent years between business and IT. The software industry has needed a new way of conceptualizing and practicing software development. Object-oriented technology arose in part as a response to this need.

Object-oriented technology provides a new approach to software development.

Object technology focuses on the creation of self-contained pieces of software. These pieces very often represent objects in the real world, such as people, places, and things. Objects are self-contained in that information about an object and the programming that operates on that information are combined within the object. An object maintains the integrity of its information by ensuring that it is changed only through the use of the programming provided by the object. All of this is done without requiring the user of a software object to know how the object does its work

**Objects are self-contained pieces of software.**

As David Taylor points out, object orientation brought more to the software development industry than a new technology; it brought an entirely new way of looking at the practice of software development. It allowed software development to move from a craft into a construction industry.

**Object technology emphasizes software reuse.**

A construction approach to software development focuses on combining objects into working systems. Whenever possible, developers reuse objects that have been developed in a previous project or purchased from a vendor. Because objects are self-contained and perform predefined sets of functions, they can be reused in many software development projects.

To continue the furniture analogy, a construction approach to software development through the use of object technology is comparable to an assembly line approach to building furniture. To support the assembly line approach, individual parts of furniture were constructed by individuals specializing in that work. This is the division of labor for which the industrial revolution is famous. Assembly line workers were responsible for constructing the finished piece of furniture from the parts produced by the specialists.

**Assembly lines emphasize efficient construction.**

In a similar way, the software development industry has begun to move toward efficiency. The construction approach to software development, facilitated by the

Object technology moves software toward efficient construction.

object-oriented emphasis on reuse, results in a better match for a construction- (or assembly line) oriented business model. In a manner similar to the way in which an assembly line partitions manufactured product into pieces, objects make possible a significant step toward partitioning software into "pieces" that can be used to construct software products. This change in approach has brought hope for relief from the software crisis.

## The Limitations of Object Technology

Object technology is limited.

Object technology has brought efficiencies to the software industry that in some inspiring cases are comparable to the efficiencies of the construction-focused, assembly line approach long ago adopted by business. This has brought hope that application development backlogs can be minimized and defects reduced.

However, just as software developers began to see the proverbial light at the end of the tunnel, significant changes occurred at two levels: the business environment, and the position of IT in the organization. The business environment became oriented toward change, and technology became embedded in organizational infrastructure. Unfortunately, object technology alone will not enable the software development industry to meet the demands posed by these shifts.

Object technology must evolve to fit a change-oriented world.

Object technology has resulted in a construction-oriented view of software more compatible with the traditional business environment than the current environment. This means it is focused on serving the relative stability of the traditional business environment. However, today's business environment is anything but stable.

Once again there is a need for greater compatibility between the software development process and the business environment it serves. Just as there was once a need for the software development industry to mature into adoption of a construction approach to match an assembly line oriented business environment, today's software processes and tools must evolve to support a change-oriented business environment.

Examples of this problem of "fit" are found in other elements of business infrastructure. The manufacturing industry provides one such example. The manufacturing processes arising from the industrial revolution often assumed a single best way to produce a product could be derived from efficiency analysis. The results were processes defined by and developed around this "one best way," which meant that the processes, once put into place, were limited in their adaptability.

As the manufacturing industry matured and grappled with the needs of a change-oriented environment, it discovered that the "best way" to manufacture a given product may change frequently, depending on factors such as current market conditions and evolving customer service requirements.

This situation demands flexible manufacturing processes that can change on very short notice, with minimal impact to quality (recall the discussion of speed, quality, and flexibility). In response, industrial engineers discovered ways to create flexible factory plans that allowed machinery and processes to be easily modified.

Similarly, traditional object technology does not serve a dynamic business environment any better than rigid manufacturing assembly lines. Object technology breaks software into manageable units in much the same way the machines and people of an assembly line break the manufacturing process into manageable units. Just as industrial engineers redesigned the tools and processes of traditional manufacturing to provide flexibility, software engineers are redesigning their tools and processes to support today's dynamic environment.

<u>Other elements of infrastructure have had the problem of "fit."</u>

<u>Static processes must be made flexible.</u>

<u>Traditional objects are relatively static, and should be made more flexible.</u>

<u>Static processes must be made flexible.</u>

However, there are elements of object technology that are resistant to this redesign effort. Therefore, even though the development process has been improved by allowing software to be "constructed" rather than "crafted," the resulting systems still lack the high degree of flexibility required by today's environment.

<u>Traditional object reuse is limited.</u>

One such element is a lack of standards for how objects interact with one another. This has limited the sharing and reuse of objects. The result is that objects are associated very closely with the languages and programs that use or create them. In fact, in the strictest sense, typical objects exist only within a single program. Therefore, although objects are self-contained, they cannot stand alone. This means that objects cannot be interchanged easily once a system has been constructed.

<u>Traditional objects are difficult to work with.</u>

Object technology also has limitations in its ability to support speed. Object-oriented programming does enable faster software development than the "build from scratch" model of the craft-oriented approach. However, the complexity of object-oriented software often limits this benefit, as expressed in the following illustration. The best object-oriented languages are difficult to learn, which, in addition to limiting the number of skilled object-oriented developers, does nothing to narrow the gap between software developers and those for whom they develop systems.

*Traditional objects have barriers to new uses and environments.*

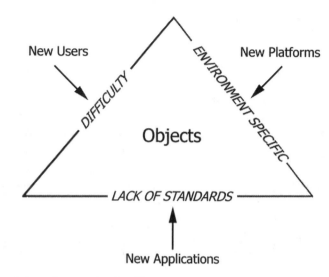

New Users    *DIFFICULTY*    *ENVIRONMENT SPECIFIC*    New Platforms

Objects

————— *LACK OF STANDARDS* —————

New Applications

Object technology is largely inaccessible to the nonprogrammer.

Consider the typical office worker's relationship to the accounting profession. The typical office worker can easily understand basic accounting concepts and can just as easily implement them in a standard spreadsheet software package without consulting an accountant. The same cannot be said of the typical office worker's relationship to the software development profession.

The office worker either delivers requirements to a software developer (who mysteriously delivers software meeting the requirement), or does nothing and receives nothing. There is no middle ground in which a typical office worker can easily understand basic object software concepts and assemble their own software without consulting a software professional. Such a situation limits the speed with which software can be created to fulfill the needs of an organization.

Object technology has been able to provide a way out of the inefficient, craft-oriented model of software development and into a construction-oriented model, but it continues to assume a relatively stable business environment. If the assumption of stability is no longer valid in today's

<u>The assumption of a stable business environment is no longer valid.</u>

## *Software Component Technology*

<u>The object model must be expanded.</u>

<u>Software components are objects that can stand alone.</u>

change-driven environment, thereby revealing the limitations of the typical object-oriented model, what can be done? The answer is that the object model must be extended (i.e., new capabilities added to the model) to overcome its limitations.

If the object model is expanded to provide a better way for objects to communicate with one another and their users, many of the limitations of current object technology will be overcome. If objects can communicate more easily with one another, they will be less tightly bound to specific operating environments, including hardware, operating systems, and even the software systems built from the objects themselves.

This freedom from limitations would make objects able to stand alone, and more interchangeable. Systems built from these objects would therefore be highly flexible. If objects communicate more easily (e.g., have intuitive interfaces) with their users, they will be much easier to use. With easier-to-use objects, systems could be developed faster, even by the typical office worker when appropriate.

Attempts to extend the object model in these two areas have resulted in the concept of *software components*, sometimes referred to as *distributed objects* or *componentware*. Software components are objects or collections of objects that stand alone. Components are independent of hardware, software, and other elements of the computing environment.

In addition, components are not inherently bound to the applications developed from them, as traditional objects are. This independence from environment and application means components can be easily mixed and matched to produce applications, and easily interchanged, even in a completed application.

## From Construction to Assembly Using Components

*An example of using components.*

As an example of software development using components, consider a specialized work management system developed by a Southern California consulting firm for one of its clients. The client required features including charting, calendar scheduling, map display, and database query.

The system was designed around the use of existing components that could be acquired from several software component providers. These components were acquired and then assembled to create the finished application. Although some customized programming was necessary to fulfill the unique requirements of the client, the use of existing components allowed the application to be developed rapidly while maintaining its flexibility.

For example, the addition of charting capability (e.g., pie charts and bar charts) to the application was a last-minute request from the client. The application was modified in a single morning by simply "plugging in" a charting component. The application developer was not required to know the complex details of how to develop charting software; rather, this knowledge was the domain of the vendor who provided the component. In turn, the vendor was not required to know or design for the applications that would incorporate the component, or to know about any future components that might be added to the client's system.

*Evidence of the power of the component approach.*

Through the use of software components, the consulting firm was able to satisfy its client by delivering an innovative, flexible, high-quality application that exceeded the client's requirements. According to members of the application development team, the achievement of these goals was directly associated with the use of a component approach to the application development.

A component approach, as represented in the following illustration, allows the software development industry to move from a construction approach to an assembly

Components move software
development to an assembly
approach.

approach. Pre-existing components are mixed and matched to create the desired application. Such an approach, with its focus on rapid assembly of interchangeable components, provides a model for software development that matches today's change-driven business environment.

*Applications can be assembled using software components.*

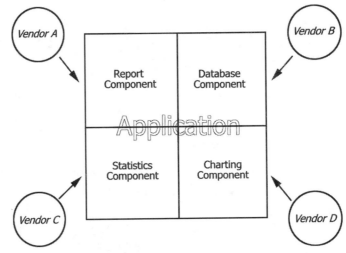

Where a craft approach focuses solely on the product, and a construction approach expands to include a concern for the efficiency of the process, an assembly approach considers the person making the product, as well as the product itself and the process. For example, returning to the analogy of furniture construction, an assembly-oriented approach to furniture does not require that the assembler possess any furniture-making skills at all beyond the basic concepts and tools.

An assembly approach
makes construction
more simple.

The assembly approach
must achieve construction
accessibility.

In fact, in some cases the assembly line and the skilled laborer working on the line are avoided altogether. The furniture company can ship a box of unassembled parts to the buyer of the furniture. The buyer can follow instructions to assemble the parts into a completed piece of furniture. This abstraction of the construction process to a point at which it becomes an activity accessible to nonexperts is one aspect of the assembly approach that software development must achieve.

First-generation component assembly tools stress a visual (i.e., graphic) approach to application development using components. In a visual development environment, such as Visual Basic or Delphi, components are dragged-and-dropped, stretched and resized, and clicked and double-clicked with a mouse. These development environments not only increase programmer productivity but are fore-runners of future visual development environments that will enable nonexperts to use components.

These visual development environments will be to the software development industry what spreadsheets were to the financial industry. They will eventually enable the typical office worker to assemble components quickly for personalized applications.

Components also support a sort of division of labor that can increase software quality. Those who are experts in a given field can reliably produce high-quality components for their given specialty. For example, Environmental Systems Research Institute (ESRI) has used its expertise in the field of geographic information systems to produce a mapping-oriented component called MapObjects.

With MapObjects, software developers who require complex map-related functions in an application, and who may not have a deep background in geography, do not need to program these functions themselves. They can acquire the component and immediately begin adding mapping functions to their application, with the knowledge that the software they are relying on has been developed by experts in the field.

Standards for how components communicate with users and with one another are being developed and implemented. These include standards such as OLE/COM and CORBA, which are discussed in detail later in this book.

*New tools allow assembly of components into applications.*

*The future of personalized applications.*

*Components increase quality through division of labor.*

*Cmponent example: ESRI's MapObjects.*

Emerging standards provide compatibility between components.

This means that standardized components can be combined without regard to which vendor created the component. Industry-wide standards will allow, for example, the developers of the work management application mentioned earlier to swap out the application's charting component for another vendor's charting component with minimal impact to the application.

Standards for components also provide flexibility for how a system is implemented. Standardized components that exist independently of their operating environments can communicate with one another across networks, on different hardware from different manufacturers, even while running different operating systems. As business needs drive frequent hardware upgrades and substitutions, a component-based application can remain stable even in a changing computing environment.

Component standards provide flexibility.

## A Component Case Study

As an illustration of the benefits of software components, consider the case of Purolator Courier Ltd. as cited in an article titled "Programming Goes Prefab," in the January 13, 1997, issue of *Information Week* magazine. As Canada's largest courier service, Purolator was struggling with the movement of a million packages per day through its processing systems. In addition to supporting the sheer volume of processing, software supporting the company's operations needed to be flexible enough to support rapidly changing requirements from clients, such as the addition of a new warehousing service.

Purolator Courier Ltd. needed fast and flexible software.

Purolator also discovered that the degree of flexibility in the company's information systems determined the speed with which Purolator could respond to client requests for changes in the shipping process. This speed of response ultimately determined whether the company would retain business or forfeit it to other shippers who could quickly

Purolator's market
share depended
on software quality.

incorporate the requested changes. "With traditional development, we couldn't deal with change," says Philippe Richard, chief IT architect at Purolator. "Taking twelve months to build something three customers asked for didn't make sense."

As if flexibility of such complex systems were not enough of a challenge, Purolator recognized that the quality and reliability of these system changes contributed significantly to customer loyalty and the long-term relationships between Purolator and its clients.

Purolator implemented
a component-based
architecture.

Purolator determined that the only way to meet these critical needs for speed, flexibility, and quality was through software components. The company implemented a component-based architecture in its information systems. According to Richard, in Purolator's new component-based architecture, "There is no such thing as a program, application, or system in our strategy. We have only components: Lego blocks." The new architecture not only allows Purolator the fast, flexible, and reliable systems development required, but positions the company for rapid expansion into as yet unforeseen business opportunities.

Purolater used software
components to meet
business demands.

Purolator is not alone in its belief that software components provide a new way of approaching software development that yields very real business benefits. TransQuest Information Solutions, an information technology subsidiary of Delta Airlines, is moving toward a component architecture to grapple with the complexities of providing technology solutions to over twenty airlines. Louis Marbel, director of enterprise application development at TransQuest, has said, "We are no longer building data processing applications. We are componentizing our business."

Likewise, when planning for future expansion of its inventory to include new media formats that provide alternatives to CDs and videotapes, Warner Music France chose to reengineer its warehousing space management application using software components. In addition to positioning Warner's systems to adapt quickly to future inventory requirements, a component approach accelerated the reengineering process. Gabriel De Fombelle, information systems director of Warner Music France, has said, "We spent a very short time developing the core of the system; perhaps a little more than a month. It would have taken us at least a year to do the same thing without prebuilt components."

These experiences of Purolator and others serve as real-world examples of how speed, quality, and flexibility are demanded from information systems functioning at the infrastructure level of an organization. They also provide real-world examples of software components meeting these demands.

## No Simple Solution

As indicated in the following illustration, the IT infrastructure of an organization is much more complex than any single technology. As a result, it would be simplistic to believe that advancements in a single technology, such as software components, can solve all problems in the software development industry. The software industry has long been desperately seeking a simple solution to its problems, but as Fred Brooks described (IEEE computer conference proceedings, 1987), "There is no single development, in either technology or management technique, that by itself promises even one order-of-magnitude improvement in productivity, in reliability, in simplicity."

*No single technology will solve all problems.*

*Software components are just one element of a new approach.*

# The IT Infrastructure

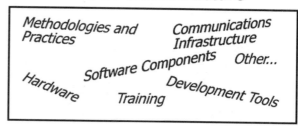

Significant progress in meeting the demands placed on IT will only come as technology planning and implementation are improved on many fronts, including quality of personnel and processes in addition to improved technology.

*Components are a key element in a new software development paradigm.*

Software components must be seen as just one element of a new assembly-oriented approach to software development. Fully realizing a new approach to software development will require improvements in many areas, such as methods for aligning component implementations with business strategy, methodologies for development of applications using components, component assembly tools, management of component libraries, and education of developers and users in component concepts and techniques. If implemented in the context of a broad approach toward the improvement of software technology, the introduction of software components may prove to be one of the most significant advances in the short history of software development.

# 2

# Software Components and GIS

The previous chapter presented a business case for software components by briefly examining the current business environment and how a component approach benefits that environment. In addition to the general business case for software components, however, there is also a unique case for software components within the geographic information systems (GIS) industry in its current state. This chapter examines why a component approach is critical to the continued growth and development of the GIS industry.

*The GIS industry needs software components.*

## A Brief on GIS

For those unfamiliar with GIS, it is worth taking a brief detour here to note the basics of the technology. GIS is, in essence, the automation of map-based information. This information, stored in any number of ways, is usually displayed as a computer-produced map. One of the unique features of GIS is its ability to relate locational map data, often referred to as spatial data, to relevant non-locational data.

*GIS automates map-based information.*

Pointing and clicking on a digital map at a point representing the location of a school, for example, might allow a GIS user to retrieve information regarding the school, such as name, level, and student population. In a typical relational database, a value such as a school name can be submitted to the database to retrieve more information regarding that school. In much the same way, GIS uses geography (in this case, the location of the school displayed on a digital map) to retrieve information from the database.

GIS uses geography as a database key.

GIS links geographic locations and information about those locations.

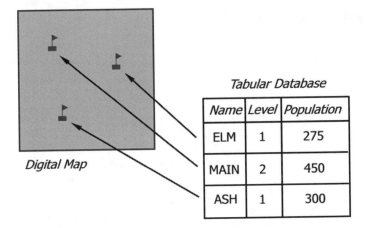

Digital Map

*Tabular Database*

| Name | Level | Population |
|------|-------|------------|
| ELM  | 1     | 275        |
| MAIN | 2     | 450        |
| ASH  | 1     | 300        |

GIS not only allows the retrieval of information through the use of geography, it also provides new methods of displaying the information. GIS software provides cartographic tools to render data in the visual form of a map. For example, a map of schools could be produced that would show each school as a point, the colors of which would indicate the school's place in a range of student population. The speed with which such maps can be produced, and the flexibility of the tools provided to create them, are making maps as easy and useful to produce as the reports and charts typically created from relational databases.

GIS displays data using geography.

Perhaps the most defining feature of GIS is its ability to perform sophisticated analysis using geography (often termed spatial analysis) that has traditionally been either impossible or extremely slow to perform. For example, a planner might use GIS to quickly determine which parcels of land fall within several environmentally sensitive areas. Without GIS, this would normally be performed by overlaying several maps on a light table and visually identifying and recording the parcels that meet the criteria.

*GIS analyzes data using geography.*

With a GIS, the digital maps are overlaid in the computer by the GIS software, and the results are displayed either as a new map or in typical report form, or both. The GIS version of this analytical operation can be performed in much less time than is required by the manual, light table process. For example, the following illustration depicts an overlay of three maps. The GIS reveals that site A is both in the "area of concern" and closer to the "site of concern" than is site B.

*An example of simple spatial analysis.*

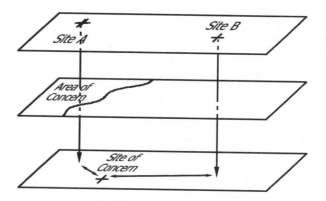

With capabilities such as these, it is no wonder that a wide variety of professionals are using GIS to record, display, and analyze geographic information. City planners are using GIS to develop and monitor urban plans. Market analysts are using GIS to predict and plan business growth. Environmentalists are using GIS to perform environmen-

A variety of professionals use GIS.

tal impact studies. Law enforcement officials are using GIS to perform crime analysis and plan the effective deployment of peace officers.

The GIS software component market is growing.

There are signs that the GIS industry is ready for component software. GIS software component products are rapidly proliferating, with such products having been released or announced by most of the leading vendors in the GIS industry. This sudden growth has been accompanied by the development of innovative applications using GIS software component products.

Drivers behind GIS industry readiness for components.

The characteristics of GIS software components have enabled applications to incorporate geographic technology where it was not previously possible, including field-based mobile computing applications and mapping applications for use by, for example, astronauts in spacecraft. This increase in vendor supply and user demand is one confirmation of the GIS industry's readiness for a component approach.

## GIS in the Information Infrastructure

Most data includes a geographic element.

There are many reasons for the timeliness of a component approach in the GIS industry, but perhaps the most fundamental is the increasing recognition of the fact that GIS has the potential to become part of the technical infrastructure of most organizations. As shown in the following illustration, various estimates indicate that 60 to 80 percent of data has a geographic element, such as addresses or geographic coordinates.

*An estimated 60 to 80 percent of data has some geographic element.*

Address

## 80%

Latitude Longitude

Census Tract

Market Area  Zip Code

Map Page

*Data's geographic element and the bottom line.*

It is important to understand that the geographic element of almost all data is at the same time omnipresent and extremely underutilized. Under these conditions, a technology such as GIS, which provides ways to take advantage of geographic information, is highly desirable across many industries and organizations. Recognizing geographic data as an underutilized, extremely valuable asset, many organizations are now implementing GIS to achieve significant business benefits.

*GIS is becoming a core technology.*

When organizations implement a GIS, they quickly find that it can provide new, useful geographic information to many facets of the organization. As GIS becomes important across the organization, it becomes part of the set of technologies comprising the IT infrastructure. If GIS becomes part of the technical infrastructure, it is subjected to the same demands on IT discussed in the previous chapter.

*GIS as a mainstream technology.*

As a result, GIS must continue to participate in the component approach with which the rest of the technology industry is attempting to address organizational demands. Software components are must be supported by the GIS industry if GIS is to continue on its path toward becoming a mainstream technology.

GIS was not always such an important technology to organizations. GIS began as a niche technology, then developed into an enterprise technology, eventually reaching its current state as a candidate for infrastructure technology. In each of these eras of its evolution, there are three specific areas by which changes in the GIS industry can be tracked: the characteristics of the GIS user base, applications of GIS technology, and the technologies provided by industry vendors.

To provide perspective on the present importance of the component approach to GIS, the rest of this chapter defines and examines the niche, enterprise, and infrastructure eras of the GIS industry, and how changes in the users, applications, and technologies at each era compare with, and progress toward, a component approach.

# GIS Evolution Toward a Component Approach

In examining GIS's evolution toward a component approach, two misconceptions must be avoided. First, although the evolution of GIS may be divided into several eras, the various eras of this evolution do not begin and end at strictly definable moments in time. The view of the evolution of GIS presented in the following sections is intended to describe the major movements in the development of the GIS industry. The changes in emphasis described in the following sections as the niche, enterprise, and infrastructure eras of GIS have occurred gradually, and overlap significantly. It is more helpful to understand the characteristics of the progression than to attempt the development of a timeline.

Second, the progression of GIS through several stages as described in the following sections does not imply that the users and approaches characteristic of a given stage cease to exist as the industry moves into a subsequent stage. For example, geocentric uses of GIS, characteristic of the early stages of GIS, still exist and provide value today. However, the proportion of total GIS use they represent is likely to diminish over time.

## The Niche Era of GIS

The niche era of GIS consisted of *geospecialists* (geographic specialists) developing *geospecific applications* using *software toolkits*. The term *niche* is used to refer to the fact that GIS technology at this time was primarily confined to the realm of geography and had not really begun to make its way into other fields. These characteristics of the niche era of GIS are shown in the following illustration, and are explained in the following sections.

*Niche era characteristics.*

|  | **Niche Era** |
|---|---|
| *Users* | **Geospecialists** |
| *Applications* | **Geospecific** |
| *Technology* | **Toolkits** |

### Users

*Users were geographic specialists.*

Geospecialist users during the niche era of GIS were, as the name implies, primarily specialists in fields closely related to geography. These included professionals—such as urban planners, geographers, and cartographers—primarily employed within regional and national land-related government agencies.

### Applications

*Applications were specific to geographic fields.*

Because users of GIS during this era were primarily geospecialists, GIS applications at this time were also typically geospecific. The applications focused on solving problems related specifically to geography, such as suitability analysis and land planning. Typical of these types of applications was the Canadian Geographic Information System (CGIS).

Implemented in 1964, and regarded as one of the first GIS implementations, CGIS was developed to store and analyze data from the Canada Land Inventory. Planning professionals used the system in order to find marginal farms by analyzing land inventory.

## Technology

It is not surprising that technology at the height of the niche era came in the form of comprehensive, sophisticated toolkits tailored for use by geographic specialists. Although the first GIS tools were small in number, highly focused, and rudimentary in function, it was not long until substantial numbers of more complex tools were created and then combined into toolkits. Technicians could pick and choose appropriate analysis tools from these toolkits to apply in a given situation.

GIS software consisted of rudimentary and, eventually, complex toolkits.

The growing amount of spatial data created in new GIS applications led to a demand for tools to maintain the data. In response to this need, a wealth of tools for maintaining spatial data were developed and added to the expanding GIS toolkits. As with the analysis tools, these maintenance tools demanded specialized knowledge to understand the processing required for spatial data, because at this point spatial data was substantially different from aspatial data in its structure and storage.

Tools for maintaining geographic data evolved.

In spite of the emergence of maintenance tools, there were few tools for managing spatial data once it was captured. Spatial data management was ad hoc at best, usually consisting of documentation of file locations and heavy use of operating system utilities.

The large, specialized applications developed with software toolkits required large amounts of computing power. Mainframes and minicomputers were the hardware platforms of choice to meet the demand for processing power. Personal computers did not exist in substantial quantities until later in this era; consequently, neither did a demand for GIS technology on the desktop.

Applications required powerful computers.

### Relationship to Information Technology Infrastructure

No substantial integration with information technology.

As represented in the following illustration, GIS applications developed during this era were typically isolated from one another, with no communication or integration between applications. In addition, these early applications had little interaction with database management systems, which were emerging as core technologies in developing IT infrastructures. IT groups within organizations were often unaware of the existence of GIS technology, except insofar as they supplied the hardware necessary to support it.

Isolated GIS implementations in the niche era.

```
                              ┌─────────────────┐
                              │ GIS Application │
               ┌─────────────────┐──────────────┘
               │ GIS Application │
  ┌────────────┴─────────────────┴──────────────┐
  │                                              │
  │              IT Infrastructure               │
  │                                              │
  └──────────────────────────────────────────────┘
```

## The Enterprise Era of GIS

Enterprise era characteristics.

The enterprise era of the GIS industry consisted of *geographic professionals* developing *integrated GIS applications* through the use of *application frameworks.* The characteristics of the enterprise era are shown in the following illustration and are explained in the following sections.

|  | **Niche Era** | **Enterprise** |
|---|---|---|
| **Users** | Geospecialists | GIS Professionals |
| **Applications** | Geospecific | Integrated |
| **Technology** | Toolkits | Frameworks |

## Users

In the enterprise era, non-GIS users began to see the advantages of GIS technology and the user base broadened as GIS spread from departments that typically used GIS to other functional divisions of organizations whose missions relied heavily on geography. This shift occurred primarily in the public sector, with groups such as planning departments, public safety agencies, and municipal utilities all developing GIS applications. The GIS user base broadened to include many more professions than in the niche era.

*User base broadened to non-geospecialists.*

To service the needs of this growing user base, a pool of GIS professionals developed. These GIS professionals were technical specialists deeply familiar with the sophisticated GIS toolkits in use at that time. They became responsible for combining tools from the toolkits into more user-friendly systems that could be deployed to non-GIS professionals.

*A workforce of GIS professionals developed.*

Advancements in GIS technology at this time allowed GIS software to run on personal computing platforms and to provide simplified subsets of sophisticated toolkits. As a result, GIS began to be accessible in limited form to less technical end users, such as managers and decision makers.

"Business geographics" applications appeared during this era. Such applications represented a significant broadening of the GIS user base to include private sector business professionals. Market analysts, real estate professionals, and delivery route planners began to use GIS with great effectiveness. The same advancements in technology that enabled public sector decision makers to gain access to GIS enabled these business users to begin productively employing the technology.

*Private business began to use GIS.*

## Applications

Those within multiple functional divisions employing GIS eventually began to see GIS as a technology that could facilitate the integration of their activities across divisional boundaries. Single divisions also began to implement GIS systems to integrate multiple business processes within the division.

This integration of activities was possible in large part because of the common element of geography found in the information used by various divisions. This integrating role of GIS technology gave rise to the concept of "enterprise GIS," with innovative applications focusing on the interconnections between departments and work processes.

An example of such an enterprise GIS in the public sector is the San Diego Geographic Information Source (SanGIS), formerly known as the Regional Urban Information Systems (RUIS) project in San Diego, California. This project was chartered to create a regional GIS system for the City of San Diego, the County of San Diego, and other municipalities in the region. Launched in 1986, the project called for participation by 28 city and county agencies, with applications ranging from environmental analysis to emergency preparedness planning.

Similar to most applications in the niche era of GIS, most business geographics applications developed during this time were standalone. Although this was in contrast to the enterprise emphasis of shared data and technology (the primary characteristic of this era), it did represent the first recognition by the larger business community of the value of GIS.

## Technology

To support a broadening user and application base, GIS technology vendors developed condensed, easier-to-use toolkits. In addition to their compactness and user friendliness, these tools focused on spatial analysis and query functions of interest to a broadening audience, including decision makers at higher levels in the corporate structure.

The new toolkits placed less emphasis on data maintenance, which still required GIS professionals and complex tools. Spatial data management tools were also developed for GIS professionals to manage an increasing amount of spatial data. However, most spatial data management tools were proprietary in nature and did little to integrate spatial data and its management with data of other types stored in relational database management systems.

In a significant departure from the toolkit concept of the niche era, the new compact toolkits were designed to serve as application frameworks. An application framework is a toolkit whose structure provides a modifiable template for creating customized applications.

ArcView, developed by ESRI, is an example of a GIS toolkit designed as an application framework. ArcView provides a set of geographic analysis and query tools that can be modified and extended using an accompanying object-oriented language called Avenue. ArcView has been customized by users and developers to create applications in fields ranging from environmental analysis to health care service planning.

The broadening user base, including decision makers and business analysts, required that GIS run on the desktop rather than in the computer room. Just in time for this demand, decreasing price/performance ratios (more power for less price) of computer hardware made the computing power required by GIS available on the desktop. Workstations and high-end personal computers thus became the platform of choice for many GIS applications.

## Relationship to Information Technology Infrastructure

*GIS applications integrated with one another.*

As previously noted, a major emphasis in application development during this era was the integration of multiple GIS applications across work processes or even multiple departments. Such integration contrasts with the isolated applications of the niche era of GIS. The possibility of integrated applications was typically taken into consideration in the enterprise era when an organization planned its IT infrastructure.

*GIS infrastructure developed parallel with information technology infrastructure.*

Staff located within the IT departments of organizations were often, though not always, influential members of enterprise GIS implementation teams. At the technical level, however, little integration existed between the tools, applications, and databases of the enterprise GIS efforts, and the tools, applications, and databases of the IT infrastructure. As represented in the following illustration, the result was a layer of GIS infrastructure conceptually resting on top of, and distinct from, the IT infrastructure of the organization.

*GIS implementations integrating with one another, but not with IT infrastructure.*

| GIS Applications |
| --- |
| IT Infrastructure |

# The Infrastructure Era of GIS

The developing infrastructure era of GIS will eventually consist of a *broad range of users*, including software engineers and solution developers, embedding GIS capabilities within a *broad range of applications* through the use of *geographic software components.* The characteristics of the infrastructure era of GIS are shown in the following illustration and are explained in the following sections.

|  | *Niche Era* | *Enterprise Era* | *Infrastructure Era* |
|---|---|---|---|
| *Users* | Geospecialists | GIS Professionals | Universal |
| *Applications* | Geospecific | Integrated | Embedded |
| *Technology* | Toolkits | Frameworks | Components |

## Users

The enterprise approach to GIS, which has integrated multiple functional groups or multiple work processes within a single group, parallels a shift in the general business environment toward process-oriented business models. In this environment, the elimination of functional boundaries and the broadening of individual job responsibilities are common. As a result, users of technology are becoming increasingly cross-disciplinary, and are solving complex problems requiring multiple technologies and multiple sources of information.

In this way, the user base of GIS is becoming much more inclusive. To become established as a mainstream technology, GIS must become accessible to a nearly universal user base. Geographic technologies must become as easily accessible as the paper-based road map atlas currently is.

Virtually anyone can use a common road atlas, regardless of their level of knowledge of geography and cartography. Similarly, geographic technologies must be accessible to anyone working in a digital environment, regardless of their level of understanding of geography or the peculiarities of GIS technology.

## Applications

Much of GIS technology
should be invisible
to the user.

The applications developed for a universal user base will typically be process-oriented applications that contain, embedded within them, the appropriate level of GIS capability. The GIS functions will be integrated with the rest of the application functions in a way transparent to the application user.

Such transparency is in keeping with a trend in the IT industry toward eliminating user-level differentiation between technologies. For example, just as the specific hardware platform required for processing should be completely invisible to the user of an application, the specific software technology required to perform a given function should also be invisible.

An example of transparent
GIS technology.

For example, a work order management application would be focused on the process it is intended to support: work order management from start to finish. The user may require a GIS function as part of the application; for example, the ability to create a map displaying work orders color-coded according to priority. When a user requests this function, the work order map is created and displayed within the same user interface that presents the non-GIS portions of the application. The user does not necessarily need to know that a specialized technology called GIS was required to perform the function.

GIS must converge with
other technologies.

Many technologies are already transforming to support the technology integration demanded by a universal IT user base. The concept of *information appliances*, which may be defined as the integration of multiple technologies to create consumer items, is one result of such transformation. Information appliances ranging from Internet-aware telephones to digital cameras have become possible as technologies have converged. The ability of GIS to converge with other technologies to provide new solutions, perhaps even at the level of information appliances, will be enabled in part by GIS software components.

## Technology

There are three primary ways in which GIS technology must transform to meet the requirements of the new users and applications of the infrastructure era. First, as the following illustration suggests, GIS technology must become *embeddable*. GIS tools should seamlessly and transparently interact with non-GIS tools within the same application. This is true in regard to the programmer, as well as the user, of the application.

*GIS technology must become embeddable.*

In the foregoing example of the work order management system, the GIS function was embedded in the general user interface of the entire application. This eliminated for the user any need to distinguish between GIS and non-GIS functions. From the programmer's perspective, the GIS component must be "embeddable" within a generic programming environment. This means that the GIS component must be usable within the programming language and interface used to develop the non-GIS portions of the application, as represented in the following illustration.

*GIS will become embedded in applications.*

### From independent...

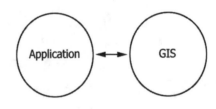

### ...to embedded.

GIS technology must also become *scaleable.* In contrast to the toolkit approach, which includes every conceivable GIS tool that might be required, GIS technology of the infrastructure era must provide a means for an application to carry with it only those GIS functions required for the application's purpose, as represented in the following illustration. The GIS tool should perform the specific functions required only when called upon to do so, without the overhead of a large toolkit of functions unnecessary to the application.

*GIS technology must become scaleable.*

*Instead of accessing an entire GIS when only a part of it is needed, applications will embed only those GIS functions they require.*

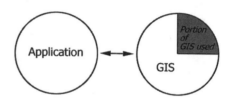

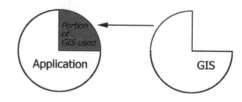

*GIS technology must become resource minimal.*

As indicated in the following illustration, GIS technology must become *resource minimal.* GIS tools must demand a minimum of computing resources, such as memory and disk storage. Future GIS applications may not have access to the resources required by typical high-end GIS toolkits.

Although it is true that hardware resources are becoming less expensive, it would be dangerous for the GIS industry to disregard resource considerations. Many new technologies, such as mobile computing and the Internet, are resource constrained for reasons other than cost. GIS tools that can fit the low resource profile of such technologies will enable the creation of innovative and important new GIS applications.

*GIS software will not require large amounts of dedicated system resources.*

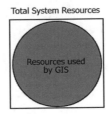

*From resource intensive...*

*...to resource minimal.*

## Relationship to Information Technology Infrastructure

GIS in the infrastructure era will be more than integrated; it will be embedded, as shown in the following illustration. The term *integrated* implies *connections* between technologies in which each technology retains its distinct identity. The term *embedded* implies a *convergence* of technologies in which the distinctions between the technologies in question are reduced to such an extent that only the identity of the result (a software application, for example) is dominant. As GIS software becomes embedded within the general application software environment, it will evolve into core technology in its own right.

<u>GIS embedded in</u>

<u>technology infrastructure.</u>

*In the infrastructure era, GIS will be embedded within the IT infrastructure.*

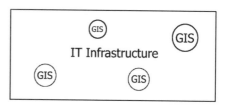

## The Critical Need for Software Components

GIS has a significant
opportunity.

The GIS industry is now at a crossroads. Information is everywhere, and it almost always includes a geographic element. Organizations are becoming increasingly aware of the benefits of data's geographic aspect. Although this provides significant opportunity for geographic technologies, it also places new demands on them to seamlessly integrate with the IT infrastructure of the organization. Software components are an important element of the fulfillment of these demands.

Components are critical
to seizing the opportunity.

The expansion of geographic technology into the fundamental requirements of IT infrastructures is paralleled by an expansion of GIS horizontally into many new application areas, such as the Internet, mobile computing, and information appliances. These new users and applications require a transformed GIS technology to service their needs. In addition, current GIS users are looking for new and improved solutions to old problems.

The result of this vertical (infrastructure) and horizontal (users and applications) expansion of GIS is an opportunity for unprecedented growth in the use of geographic technologies. Whether the GIS industry seizes this opportunity will depend to a large degree on whether it pursues the advances made possible by software components.

# 3

# Objects and Components: What's the Difference?

Objects and components are similar but have individual characteristics.

Because the component software model is an extension of the object-oriented model, components and objects share many characteristics. Despite these similarities, traditional objects differ from software components in many ways. Their differences highlight the limitations of traditional objects and help clarify the unique contributions of software components.

This chapter compares traditional objects and software components to examine their similarities and differences, and discusses the characteristics component technology will most likely develop. This chapter lays a foundation for the discussion in Chapter 4 of various models for the technical architecture of software components.

Clarifying terms.

It is important at the outset to clarify the terms *component*, *object*, and *traditional object* as they are used in this book. Fundamentally, components are based on an object model. In this sense, components are objects. However, components extend the traditional object model to produce "super-objects" with an autonomous nature.

In contrast to typical objects, components can stand alone, meaning that they are not bound by the compiling process to the applications developed from them. Also in contrast to traditional objects, they operate independently from specific hardware, software, and language environments. Because of these characteristics, components are sometimes referred to as "distributed objects." The following illustrations show how the environmental dependencies of traditional objects contrast with the independence of components.

*Components are objects that can stand alone.*

*Traditional objects are bound by their environment.*

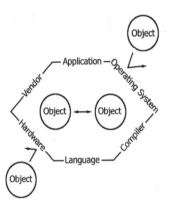

*Components can span environments.*

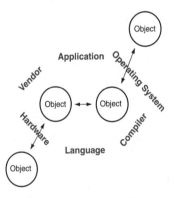

The term component is used to describe these "super objects" and stresses their essential nature from the perspective of application development. Components are the constituent parts ("component" pieces) of an application, and can be added and removed using component software development tools. Unless the context indicates otherwise, which it often does, the term object is used in this book to refer to objects in general, whether component based or not. The term traditional object is used to distinguish objects that are not components

## *A Brief Review of Objects*

To understand components, you must first understand objects. For those new to object technology, or those who need a brief refresher, this section reviews the most basic object-oriented programming concepts.

## Objects, Data, and Methods

*An object is a "package" of related data and procedures.*

There are many definitions of a software object, but perhaps the most succinct is provided by David Taylor. "An *object*," Taylor says, "is a software package that contains a collection of related procedures and data." This means that an object represents a "thing," contains information associated with the "thing," and can perform specific actions.

For example, imagine you wanted to represent an airplane using a software object. The plane would have associated information, such as weight, number of engines, and top speed. These associated pieces of information are often referred to as *object variables* or *properties*. They reveal information about the object itself. Some properties are made available to programmers and other users of the object. These properties are called *public properties*.

*Related data are called object variables.*

Other properties may be referenced only internally by the object in the course of doing its work. These cannot be changed by the user of the object, and may not even be visible to the user. These are called *private properties*.

Variables are very common in the non-object programming world. However, in the object programming world, object variables, or properties, can only be accessed by

Related procedures

are called methods.

functions (called *methods*) that are defined as part of the object. These are the "related procedures" of Taylor's definition. In the example of the airplane object, methods might be provided to allow a programmer to change the weight or the number of engines of the aircraft to correspond to the airplane the object represents.

These methods are invoked by issuing requests, also known as *messages*, to the object. Just like properties, methods can be public (accessible to users of the object) or private (used only by the object itself in its internal operations). Distinctions among methods, properties, and messages are represented by the following illustration.

*An example of object methods, properties, and messages.*

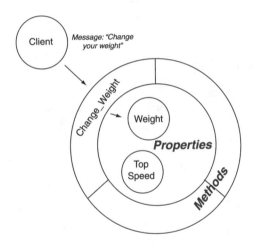

## Encapsulation, Polymorphism, and Inheritance

Encapsulation, polymorphism, and inheritance are three key concepts of object-oriented programming, and are defining elements of the object-oriented model of software development.

Using the methods and properties of an object to maintain the integrity of the object's data is called *encapsula-*

*tion*. The object specifies what data can be accessed or changed (public versus private variables), and exactly how it must be accessed and changed (through what methods). This ensures the usefulness of the object because it hides from the user, who only cares that the object *does* its work, the complex details of specifically *how* the object works. Encapsulation allows an object to be self-contained. It is the reason objects are often described as performing an action autonomously, as when we say that "the graphic object drew itself."

**Encapsulation keeps object details hidden.**

Different objects may use the same method to perform the same operation, but each object's method may be implemented differently. This is referred to as *polymorphism*, referring to multiple forms of method. For example, two different word processors may each have "spelling" objects that implement "check_spelling" methods. The two development teams creating the two spelling objects may implement their spell-check methods quite differently. However, anyone using the objects simply calls the "check_spelling" method in either case. The methods and results are similar for the user, but the internal workings of the methods differ.

**Polymorphism allows a method to be implemented in multiple ways.**

In the previous example of an airplane object, a specific airplane belongs to the general category of things called "airplanes." Airplanes can be categorized this way because they share characteristics—such as weight, number of engines, and top speed—common to every member of the category.

**Classes are categories of similar things.**

In object technology, such categories of things are implemented as *classes*. A class includes definitions for the properties and methods common to all members of the class. A class named "airplane" may include definitions for properties representing the characteristics of weight, number of engines, and top speed that are common to all airplanes, and therefore to all members of the airplane class.

By defining the common characteristics of its members, a class serves as a template for creating objects that represent actual occurrences of class members. An object representing a specific airplane—with specific values for weight, number of engines, and top speed—can be created using the airplane class as a template. This specific occurrence of the airplane class is called an *instance* of the class. The instance is what is usually referred to when we use the term *object*.

<u>Objects are instances</u>

<u>of a class.</u>

To continue the example, the category of things called airplanes belongs to the more general category of things called "vehicles." Other categories of things, such as automobiles, belong to the "super-category" of "vehicles." In object technology, the "vehicle" category could be implemented as a *superclass*. This superclass may have common methods and properties, such as weight and top speed, it shares with all of its *subclasses,* which, in this case, are "airplanes" and "automobiles."

<u>Subclasses inherit</u>

<u>from superclasses.</u>

The subclass may also add new methods and properties to the collection of methods and properties it has received from its superclass. For example, "number of engines" is not likely to be a property of the automobile class; therefore, it is not defined as part of the superclass "vehicles." The ability of subclasses to receive methods and properties from their superclasses is called *inheritance,* which is represented in the illustration that follows.

Notice that the properties "weight" and "top speed" are inherited from the "vehicle" class by the "airplane" class. Notice also that the methods "Change_Weight" and "Change_TopSpeed" are also inherited. In addition to inheriting properties and methods, the "airplane" class defines the new property "# of engines" and the new method "Change_NumEngines."

*Classes can inherit from other classes to produce object instances.*

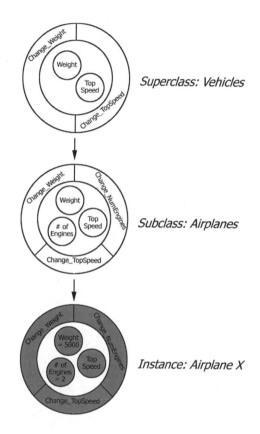

Superclass: Vehicles

Subclass: Airplanes

Instance: Airplane X

## How Objects and Components Differ

*Components are objects that can stand alone.*

The fundamental difference between traditional objects and components is that traditional objects are significantly dependent on their environment, whereas components are not. As described at the beginning of this chapter, components are essentially objects that can stand alone and operate independently of their hardware, software, vendor, and language environments. In addition, they are not inherently bound by the compiling process to the applications developed from them, as are traditional objects.

When freed from such environmental dependencies, objects can be easily removed from an application and replaced with others, communicate with other objects across networks and on different platforms, and interact with other, unknown components written by different vendors in different languages. This independence is accomplished by adding elements to the traditional object model that more clearly define the separation between objects and their technological environments. In preparation for discussing in greater detail the differences between traditional objects and components, the following sections briefly review the origins of software objects, the problems they are intended to solve, and their limitations in providing solutions to these problems.

## Defining a Unit of Software

Because the object model defines objects in a way that makes them self-contained units of software, object proponents expected that this self-containment would lead to a clean line of separation between the object and its environment. The clearer the distinction between an object and its environment, the more environment-independent the object can become, thereby making the object more manageable, interchangeable, and reusable. This need to create units of software that are more flexible is not a new phenomenon; it has origins deep in the history of software development.

*Objects evolved from the need to define manageable units of software.*

Prior to the advent of object-oriented programming, software development typically resulted in monolithic applications. The distinct characteristic of monolithic applications is the existence of heavy internal dependencies within the application's program code that prevent it from being partitioned into manageable, reusable pieces. As the size of applications grew, it became apparent that program code must be divided into manageable units. The inventions of subroutines and modular programming in the 1950s and 1960s were significant attempts to subdivide monolithic applications in this way.

*The first software was monolithic.*

Although the advances of modular programming did introduce some measure of manageability, the way in which modules were used often resulted in arbitrary divisions of program code, as defined by the particular programmer developing the application. The introduction of structured analysis and design techniques in the 1970s and 1980s provided some consistent guidelines for defining divisions of program code.

**Structured analysis and design helped partition software.**

With the introduction of these techniques, the issue of the division of software applications into manageable units was considered in the context of user requirements and design, rather than being addressed only as regarded program code. This made the divisions less arbitrary and more consistent across applications.

**Subroutines did not reflect the real world.**

However, the methods used to perform this division were somewhat artificial and were typically implemented as procedure-driven subroutines. The industry recognized that methods for partitioning applications must be improved to better reflect the real world.

The concept of object-oriented programming originated nearly three decades ago, but it was popularized in the 1980s and 1990s by two concurrent trends. First, system designers began to realize that objects provide a natural way of analyzing the real world, which itself consists of objects. Second, object-oriented languages such as C++ provided a means of implementing those objects in software.

**Objects provide a consistent view of a system throughout the development cycle.**

The combination of new methods of both designing and implementing systems using objects provided a consistent, real-world-based method of defining and partitioning the application at all phases of the software development cycle, from analysis through implementation. This was an important factor in causing the object-oriented approach to take hold.

Objects can reflect
the real world.

The concept of objects appeared to finally clarify how a unit of software and its interaction with its environment should be defined. Objects provided the clear distinction between units of software required to provide the foundation for manageable, interchangeable, and reusable units of software. Because objects reflected the "real world," it was anticipated that when a real-world problem with real-world objects changed, the corresponding objects in a software application could be changed accordingly. For the same reasons, objects were expected to be easily reusable, requiring little or no change to the original programming.

One might expect that software reuse could go a long way toward meeting the demands of today's business environment described in Chapter 1: speed, flexibility, and quality. As regards speed and flexibility, reused software should allow a programmer to develop software faster. Because the reused software is an object, its self-contained nature should make it easy to reuse and even replace when necessary, making applications more flexible.

Self-containment
(encapsulation) is not
enough to facilitate reuse.

As regards quality, if an object is developed and thoroughly tested at least once, the quality of that object is consistent for every successive project that reuses the object. Unfortunately, the achievement of reuse has been limited in actual practice.

Although the self-contained nature of objects provides the basis for reusability, traditional objects are still heavily dependent on environmental aspects such as hardware, operating systems, and programming languages. If objects cannot span these technology environments, their reusability is limited to the environment within which they exist The rapid proliferation of multiple hardware, software, and network environments in recent years, and the requirement that these environments work together, has caused the environmental dependencies of traditional objects to be a serious limitation.

## Something's Missing in the Object Model

<u>Traditional objects lack a consistent interface.</u>

The traditional object model lacks two important elements required to achieve independence from particular technology environments. First, an object must have a means of defining to its users (which include other objects as well as people) exactly what tasks the object can do and how it can be asked to do them. This definition must be consistently understandable across various technology environments and by all other objects with which the object must interact. This is the need for a consistent object *interface*.

<u>Traditional objects lack a common infrastructure.</u>

Second, once a requesting object knows that another object can do what it is requesting and knows how to ask that object to do it, the requesting object must have a means of communicating with the other object. This communication must be common across technology environments such as hardware, software, and networks. This is the need for a common object *infrastructure*.

<u>An interface is like a directory of services.</u>

The critical need for consistent interfaces and a common infrastructure is paralleled in the everyday business environment of service providers and potential clients. One way in which service providers make their presence known to potential clients is through a public telephone directory.

The interface required in an improved object model is analogous to the entries in a telephone directory. An entry in a telephone directory is a standard way for a business to state the services it provides (what it can do) and its telephone number (how it can be asked to do it). It is consistently understandable by any potential client referencing the telephone directory. Objects require the same type of means of informing potential users of what they do and how they can be asked to do it.

<u>An infrastructure is like a communications system.</u>

In the business provider/client analogy, it is not enough to have a telephone directory. The directory helps a client identify a service provider and instructs a client how to reach the service provider (via a telephone number). A mechanism must also be provided to allow the interaction between client and provider to occur.

This can be provided by a telephone system with cables, transmitters, and receivers. Such a system is required if a potential client is to actually interact with the listed service provider. Communication can occur over telephone lines between service provider and client even if they use different telephone companies.

Communication between a service provider and a client spans their technology environments. A standard specification for how telephone signals travel over the telephone infrastructure allows the service provider and client to communicate, even if they are using different telephone hardware. Objects need a similar infrastructure specification to use in their communications with one another. The following illustration shows access to services enabled by a mutual interface and infrastructure.

*An interface and infrastructure enable access to services.*

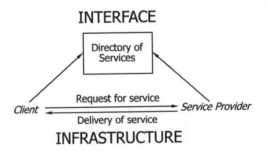

**INTERFACE**

Directory of Services

Request for service
*Client* ⟶ *Service Provider*
Delivery of service

**INFRASTRUCTURE**

<u>The component model includes interface and infrastructure.</u>

This lack of a consistent interface and a common infrastructure in the traditional object model is the primary reason for traditional object technology's limited attainment of the goal of reuse. It is also the point of departure of components from traditional objects.

A component approach extends the traditional object model to include both an interface and an infrastructure. This results in objects that are easy to use and reuse across languages, hardware, networks, and operating systems. This independence from technology environments provides more effective interaction between objects and their users, whether humans or other objects.

# Moving from Objects to Components

The following sections discuss object interfaces, object infrastructures, and the changing software world. These are three major areas to consider when assessing the technology and forces behind the trend of movement from object technology to component technology and application.

## Object Interfaces

*Interfaces tell users what an object can do.*

When discussing object interfaces, the term *user* is defined loosely. The user of an object may be a software developer, or the user may be another object. In either case, the goal of defining an interface specification is to provide a technology-independent means of defining what tasks an object performs and how an object should be asked to perform those tasks.

*An interface is an object/user contract.*

Interfaces are often described as contracts between a user and an object. The object fulfills its part of the contract by making known through the interface exactly what it can do and how a user can request a specific action. The user fulfills its part of the contract by agreeing to work with the object via the terms described in the interface (e.g., accessing the object only through its defined interface).

Because the traditional object model does not have a standard specification for interfaces between objects and their users, most object implementations do not support the sort of cross-environment user/object contract described.

Interfaces span location and environment.

For example, IPC (interprocess communication) is a term used to describe how one process or application can communicate with another process or application on a single computer. If these processes are viewed as objects, IPC can be described as an interface between objects on the same computer.

RPC (remote procedure call) is a term used to describe the way in which one process or application can communicate with another process or application on a remote computer. Implementing IPC can be very different from implementing RPC. This adds complexity when an application must be built using a combination of local and remote objects. In a better world, the world of components, a standard interface would support both local and remote object references.

Interfaces hide details of object communication.

In addition to providing a way of addressing objects in the same manner regardless of their location or environment, interfaces also mean that components adhering to a standard interface can communicate with a minimum of programmer intervention. Consider the example of a programmer using a component-aware development environment such as Visual Basic, Delphi, or PowerBuilder to create an application wherein a button, when clicked, will display a report.

In such an application, the programmer first adds a button object (created using a button component) and a report object (created using a report component) to a form (window) object. The programmer can then specify that a mouse click on the button will display a report in the form window. The programmer simply directs the button object to request a report from the report object, and directs the report object to display the resulting report in the form window. The programmer simply tells the objects what to do, and the objects' interfaces handle the complex details of how to communicate with one another to process the request.

Interfaces also enable significant improvements in the tools programmers use to work with objects. Visual development environments (VDEs), sometimes called "integrated development environments" (IDEs), are development environments that emphasize assembling components by using a graphic user interface (GUI) rather than writing endless lines of code. A programmer can add a new component to the IDE at any time. These components are dragged and dropped into "containers." A container serves as a common area in which components present themselves as a coherent application and communicate with one another.

<u>Interfaces enable a focus on assembling objects.</u>

To generate a report using a specific database, you might add a report component to a form (the container) and then drag a database component into the form container and onto the report component. A component approach allows objects to be added, mixed, and matched easily because component-based objects can instantly understand the capabilities of any other component-based object in the container. This is accomplished through interfaces.

<u>Containers facilitate object interfacing.</u>

In summary, interfaces provide at least three important benefits. First, they provide a standard way for objects to understand the capabilities of one another, regardless of their location or environment. Second, they free the programmer from knowing the details of how objects communicate. (Objects have the ability to determine this independently by referring to the interfaces of other objects.) Third, they make possible productive and easy-to-use visual development environments.

<u>Interfaces make objects easier to use.</u>

## Object Infrastructures

The second element missing from the traditional object model is an infrastructure by which objects actually accomplish their communication with one another. Spe-

Objects need
a standard infrastructure.

cifically, traditional object technology does not provide a means of cooperation between objects written by different companies, in different languages, and running on different hardware, with different software, on different machines.

The addition in the component model of a standard interface allowing objects to publish and read one another's methods is one step toward solving this problem. However, for interfaces to be useful there must also be a standard way for objects to actually send their communications to one another.

Objects communicate
via infrastructures.

To return to the analogy of the telephone infrastructure, interfaces provide us with a "directory of services" for what an object can do and tell us "where to call" in order to request its services, but there must also be some means of actually "placing the call" to the object. Object communication infrastructures fill this need.

Traditional objects
communicate within
language of origin.

Because the object model has no infrastructure element, traditional objects are limited in their communication with one another. For example, traditional objects implemented in a language such as C++ are reusable and can communicate with other C++ objects. However, this usually only works when the object has been compiled using the same compiler used for the application accessing the object.

To allow the programmer to reuse objects, object library providers often supply the source code for the objects to allow for recompiling when necessary. Also, object exchange is cumbersome at best. If you have previously used object A in your application and now want to replace it with object B, you must recompile your application with object B. Who wants to spend all of that time compiling when the goal is to develop solutions?

Because the mechanism for object communication in the traditional object model is software and hardware dependent, traditional objects are not interoperable. An interoperable object must be able to communicate across language, operating system, hardware, and vendor boundaries. Language-bound objects cannot accomplish this because objects created in one object-oriented language are not generally reusable in another object-oriented language. In addition, they communicate with great difficulty, if at all, across operating systems and hardware platforms, and almost never across vendor boundaries.

*Traditional objects communicate within environment of origin.*

The close relationship between objects and their languages, operating systems, hardware environments, and manufacturers results in a very narrow scope of interaction between objects. Revisiting our telephone directory analogy, it is as if a client can only place a telephone call to a service provider if both are using the same model of telephone, made by the same manufacturer, and connected to lines supplied by the same phone company. Business activity in such a world would certainly be limited, just as object interactions in software have been limited.

*Language and environment restrict object interaction.*

A well-implemented telephone infrastructure allows telephones made by different manufacturers to communicate over lines owned and operated by different companies. This is done by developing standard specifications for how communications occur. Any product developed in adherence to the standard will interact successfully with other existing products. A specification for a standard object infrastructure provides just such a solution for inter-object communication.

*Unrestricted communication requires standards.*

A successfully specified and implemented object infrastructure creates a component world that looks much different from the language- and machine-bound world of traditional objects. In the component world, objects use an object infrastructure to work together regardless of the language in which they were developed, the operating system on which they run, or the machines on which they reside.

*An infrastructure is language and environment independent.*

An important element of an inter-object infrastructure is the concept of "binary reuse." This means that components exist as pre-compiled binary objects. The objects do not interact at the source code level, in contrast to traditional objects, which must be compiled together in the same application. Because there is no need for source code or recompiling, an object can be added to an application and be expected to instantly interact with any other object that uses the same infrastructure. In a world of binary reuse, applications can be assembled from objects acquired from vastly different sources.

<u>An object infrastructure</u>

<u>supports binary reuse.</u>

Consider the example of the component-based work management system described in Chapter 1. It included a charting component and a map display component. The application was built using a component approach based on a common inter-object communication infrastructure. Because of the underlying component infrastructure, the charting component, which was produced by vendor A, can be easily removed from the application and replaced by a charting component from vendor B.

<u>Binary reuse increases</u>

<u>compatibility of objects.</u>

No matter which charting component is plugged into the application, that charting component will be able to "talk to" the existing map display component, produced by vendor X, and vice versa. The programmers from vendor B who created the charting component may have never imagined that it would interact with a map display component from vendor X, but if they develop the component using the common infrastructure and interfaces of the component model, the application will function without fail.

## The Changing Software World

As components begin to dominate, the world of software will experience dramatic changes from the perspective of both users and developers. The following are a few of the probable changes that will take place in the near and distant future.

❏ Interoperable components with very few operating limits will produce a mix-and-match world in which the number of possible solutions to a problem is as large as the number of possible combinations of components.

❏ The ability of components to communicate with one another across a network will provide new possibilities for distributed applications, improved use of hardware resources, and consequently better return on hardware investments.

❏ The large off-the-shelf applications that currently exist, such as spreadsheets and word processors, will become collections of components. This will eliminate the "one size fits all" application and allow customization for specific situations.

❏ The interchangeable nature of components will result in unprecedented flexibility, allowing software to be quickly adapted to changes in business.

❏ The focused nature of components will support a division of labor in which experts will be able to focus on creating components in their area of expertise, such as in GIS, without loss of component usability by nonexperts.

❏ "Throw-away" software will become possible. Applications will be quickly assembled to seize momentary business opportunities, and then disassembled and reused when no longer required.

*Components will make software development accessible.*

Perhaps the most fundamental change will be the increased accessibility of software development. Creating software by assembling components rather than by complex programming makes software development accessible to many more people. This accessibility is already giving rise to a new type of software developer, known as a "solution developer."

Solution developers are generally closer to the business processes being addressed by their efforts than are traditional software developers. Eventually, many solution developers will be business professionals, not programmers, who are part of the business process being addressed. Such a trend will provide an entirely new set of opportunities for software development in the business environment.

# 4

# Models for a Component Architecture

Understanding
OLE/COM and CORBA.

Software components, including geographic software components, must be based on an underlying component model that provides support for the interfaces and infrastructure discussed in the previous chapter. Microsoft's OLE/COM (Object Linking and Embedding/Component Object Model) and the Object Management Group's CORBA (Common Object Request Broker Architecture) model are currently the two dominant models for software components.

Managers and users of components should be familiar with the basic concepts of these models to provide a context for making decisions and understanding products in the world of software components. This chapter explains these basic concepts, as well as increasingly common component terms such as OLE, COM, CORBA, OCX, ActiveX, OLE automation, and OpenDoc.

## The Need for a Standard Component Model

The previous chapter discussed how the traditional object model must be expanded to include an interface and infrastructure. However, before it is possible to implement software components as actual products, specifications for standardized component interfaces and infrastructures must be developed.

Specifications provide a sort of requirements blueprint to which the producers of component products can adhere. Although vendors may implement their products differently, adhering to a standard specification ensures that their components and component-related products will operate with others built to the same specification. Several groups have begun to develop specifications for a standard component model, two of which have emerged as the strongest forces in these efforts: Microsoft with OLE/COM, and the Object Management Group (OMG) with CORBA.

*Standard specifications provide interoperability.*

## Microsoft and the OLE/COM Model

OLE/COM is Microsoft's vision of the component world. As you might expect, OLE and COM enhance the traditional object model by providing the all-important component infrastructure and interfaces.

## A Brief History of OLE/COM

OLE and COM evolved as Microsoft refined the capabilities of its applications in the graphical user interface environment. Early iterations of Windows applications made use of the clipboard to cut and paste content from one application to another. Cutting and pasting was the first step in sharing information between applications.

*OLE/COM began with cutting and pasting.*

When the content being "cut and pasted" grew to include various types of data—such as text, tables, and graphics—it became natural to think of them as objects. When these

Compound documents contain objects.

Dynamic data exchange provided dynamic objects.

Intelligence moved to objects with OLE.

objects were combined, the result was referred to as a "compound document." For example, pasting a graphic from another application into your text document by using the clipboard would result in a new type of document: a compound document containing a graphic object and a text object.

The problem with the clipboard approach was that once an object was pasted into a compound document, it became static. The object could not be easily modified without more cutting and pasting. To overcome this limitation, Microsoft introduced dynamic data exchange (DDE) in the late 1980s.

DDE allowed applications to communicate with objects residing in other applications. For example, a spreadsheet application could now be instructed by a word processing application, via DDE, to update the relevant spreadsheet object. DDE provided a more dynamic object interaction between applications, but its capabilities were limited and required a measure of programming expertise.

In 1991, Microsoft introduced OLE 1.0., which allowed objects in a compound document to be either "linked," which meant the object would look to the original data as its source, or "embedded," meaning the object would bring a copy of the data along with it into the compound document. Hence the acronym OLE, object linking and embedding.

With the advent of OLE, the objects within a compound document were considered components for the first time. The application used to create the compound document became recognized as a "container" of the objects, and the objects themselves could perform actions. This was a move toward the standalone nature of components, but the objects still relied on the support of their applications of origin to accomplish an action. With OLE, the balance of intelligence began to move toward the object rather than the container—an important step toward component technology.

Seeing the need for a standard infrastructure to support the communication required by these components, Microsoft introduced OLE 2.0, which included the Component Object Model (COM). COM is designed to support inter-object communication, and therefore provide the foundation for the de-velopment of software components. Microsoft has since dropped the version number of OLE, but it will continue to evolve. This is Microsoft's recognition that OLE/COM has become a component architecture that should support various implementations, even perhaps by vendors other than Microsoft.

*COM supplied an infrastructure.*

## COM as the Object Infrastructure

In the OLE/COM model, COM provides the common inter-object communication infrastructure critical to a component model. COM is what is commonly referred to in the component world as an object request broker (ORB). An ORB can be conceptualized as a software "bus," through which the communications between objects travel, as represented in the following illustration. An ORB's operation is similar to the way in which a hardware bus within a computer allows various hardware components, such as memory and the processor, to communicate with one another.

*COM is an object request broker.*

*Objects can communicate via the COM object request broker (ORB).*

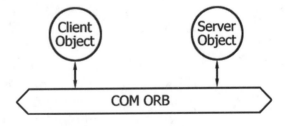

It is important to note that COM supports only communication between objects residing on a single computer. In 1996, Microsoft began to implement the Distributed Component Object Model (DCOM) to support inter-object communication between multiple computers. This support for objects on the network is critical to the success of independent components in today's distributed computing world. DCOM is implemented as an extension of COM to allow objects implemented for COM to continue functioning without failures in communication.

DCOM works by first accepting a client's request for action. COM then locates a server object to fulfill the request and passes pointers to the server object back to the client object. The client object can then communicate with the server object by using the pointers, and from that point on the two objects communicate with each other via the COM ORB.

The ORB functions as a request broker, assisting objects in establishing and carrying out communication with one another. ORBs are designed to be implemented across technology environments such as language, operating system, hardware, and vendor. These two characteristics, request brokering and cross-environment support, allow an ORB to make complex inter-object communications essentially transparent to application developers.

Independence from operating environment is, of course, dependent on the number of operating environments an ORB supports. Microsoft currently supports COM in its Windows and Windows NT operating systems, as well as on Macintosh. Several third-party vendors are supporting COM on other operating systems. DCOM extends this environment transparency to the network as well, making the location (local or remote computer) inconsequential to the object user.

*DCOM supports networked objects.*

*COM keeps object communication hidden.*

*Operating system support is growing.*

## COM Objects and Interfaces

Objects use interfaces to publish their capabilities.

The purpose of ORB is to allow objects to communicate with one another. However, simply throwing many objects together and allowing them to communicate is worthless unless the communication is meaningful. Objects that provide services must be able to identify what services they can provide and how a request for service can be made. Objects that require a service must be able to discover this information in order to begin a meaningful exchange.

For example, if your car needs repair there is no sense in taking it to your accountant. You need to find a mechanic. If you can locate a mechanic through an advertisement of services that includes an address, you can take the car to the mechanic and receive the benefit of the mechanic's services. Similarly, objects must be able to locate other objects that can supply the appropriate services. This "publication of services" to enable meaningful communication is achieved in the OLE/COM model through the use of object interfaces.

Objects use interfaces to determine the capabilities of other objects.

In the language of COM, the term *interface* is defined a bit more narrowly than it has been in previous chapters. In previous chapters, the broad definition of a component interface has referred to all elements necessary for an object to notify clients of its services and how to access those services. This broad definition embraces the more specific COM interface, as well as several other related elements such as type libraries and the Interface Definition Language (IDL), which is discussed in material that follows.

**COM interfaces are groups of methods.**

A *COM interface* refers to a group of related methods, or functions, an object can perform. A *COM object* may be defined as an object that supports one or more interfaces, which are determined by the object's class. Recall that a class serves as an "object category," defining the functions and attributes available in an object instance of that class.

**COM servers provide objects.**

COM servers are implementations—in the form of a dynamic link library (DLL) or an executable program (EXE)—that make available one or more object classes. A COM server must include a class factory that can create objects from the classes contained in the server. COM servers provide the basis for producing OLE/COM components, often in the form of ActiveX controls (formerly known as OLE controls or OCXs) that can be "plugged into" an application to provide multiple objects.

For example, a charting component may provide a chart class to allow an application developer the ability to create a chart object using values from a specified database. Methods in a graphical interface allow the programmer to draw the charts for display, and methods in a statistics interface allow the programmer to produce statistics for the data relevant to the charts.

In short, a COM server includes one or more COM object classes, which support one or more COM interfaces, which support one or more methods. These relationships are shown in the following illustration. COM objects created from the server's classes will support the interfaces and methods specified by their classes.

*COM servers, classes, and interfaces.*

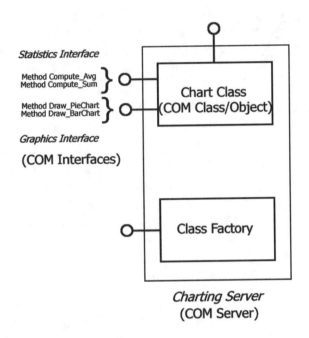

Statistics Interface

Method Compute_Avg
Method Compute_Sum

Method Draw_PieChart
Method Draw_BarChart

Graphics Interface

(COM Interfaces)

Chart Class
(COM Class/Object)

Class Factory

*Charting Server*
(COM Server)

At this point it is important to understand the distinction between in-process and out-of-process servers. In-process servers execute in the same process and therefore the same address space in memory as the client, and are generally implemented as DLLs. Out-of-process servers execute as separate processes and therefore are often implemented as EXEs.

In-process servers provide efficiency, because a call to an in-process server is within the process and is therefore essentially no different from a call to a function within the client object itself. Out-of-process servers require some traveling over the system infrastructure, and are therefore less efficient. Out-of-process servers are important in that they allow objects to reside on different machines (DCOM), yet communicate with each other. They also can provide stability, in that failure in an in-process server can bring down the entire process in which it runs, whereas an out-of-process server will bring down only its own process. In-process servers are important in a discus-

*In-process servers execute in the client's process.*

*Out-of-process servers execute in a separate process.*

sion of components because ActiveX controls, which are the OLE implementation of components, are implemented as in-process servers.

## Type Libraries and IDL: The Interface Directory Service

Server objects use a COM interface to display methods that are accessible by client objects. But how does a client object discover the interface of a server object in the first place? This capability is provided in COM through type libraries and COM's Interface Definition Language (IDL) for creating standardized interface definitions.

### Type Libraries

In COM, a type library is the "directory" wherein objects make known to other objects what actions they can perform and how they can be asked to perform those actions. In the traditional programming world, function declarations are used for this purpose. A programmer can review available functions, choose the appropriate one, and develop a program to call the function according to the specification in the declaration.

*Interfaces are published in type libraries.*

In a component world, however, objects must be able to perform this activity without intervention by a programmer. They must be able to discover, interpret, and call the methods (functions) of other objects. They accomplish this using a type library that contains a directory of the object's COM interfaces, methods, and declarations.

Once an object accesses, via the ORB, a type library listing the services of another object, its next challenge is to understand the method declarations it finds there. In the real world, businesses publish their services and telephone numbers in a telephone directory, using a language that will be understood by the greatest number of potential clients.

*Interface definitions must be understandable to any object.*

However, suppose each business used a different language to describe its services, resulting in a telephone directory with entries in English, Spanish, German, Greek, Japanese, and a variety of other languages. Client interaction with a given service provider would be limited to those clients that spoke the language of the service provider's directory listing. To avoid the same limiting situation with objects, the OLE/COM component model provides a common language for describing interfaces: the Interface Definition Language (IDL).

## Interface Definition Language

**IDL is a standard language for interfaces.**

Although COM does not specify what language must be used to write the interface definition, IDL is supported and recommended by COM as a standard. It is important to note that the actual object and its methods are programmed in the developer's language of choice, not in IDL.

An object developer creates an object's interfaces and methods as an activity distinct from the actual development of the object. The interface definition is created using IDL, with the IDL then compiled into a type library to be accessed by other objects. In this way, IDL is used as a standard language for making the methods of an object known to all other objects on the ORB.

**IDL separates the interface from the object.**

Separating the definition of an object's interfaces from the internal workings of the object itself is an important issue in the world of components, and is referred to as "separating the interface from the implementation." This means that object developers should be free to implement the object and its methods in the language and style of their choice.

The developer should also have complete assurance that such freedom will not limit the ability of other objects to access the developed object. Such freedom is possible because the actual interface definitions stored in a type library are created independently of the object's internal programming. The following illustration shows the relationship between type libraries and client objects.

*Type libraries provide client objects with the information required to call server object interfaces and methods.*

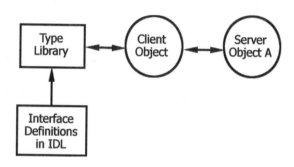

Type libraries in COM support a concept known as "late binding." This means that if new components are added while an application is running, the application can discover the new component (object) and its interfaces, and dynamically build calls to the object as required. This is common in the use of ActiveX controls, and frees the client at compile time from having to know exactly what methods it will use to perform its operation. The object knows it will eventually need a task performed on its behalf, but is not required to know prior to the performance of the task who it will ask to do the work, nor does it need to ever know how that work gets done; it simply gets results.

*COM objects support*

*late binding.*

## An Application Framework: Object Linking and Embedding

*OLE is an*

*application framework.*

OLE is an application framework that is conceptually "layered" on top of COM. The ultimate goal of a component infrastructure is to enable users to easily assemble objects into working applications. OLE provides a structure within which this assembly can be performed. Remember that OLE was originally developed to support the concept of "compound documents."

Application frameworks

enable component

assembly.

OLE and COM are often confused, largely because many of Microsoft's initial component development efforts were delivered in OLE, which was developed prior to the existence of the COM infrastructure. Now, many services that originated in OLE are being moved to the COM infrastructure, where they more appropriately belong.

However, OLE and COM are functionally distinguished by the fact that COM provides an inter-object communications infrastructure, whereas OLE provides an application framework. Object infrastructures such as COM operate at the system level to provide inter-object communication. Application frameworks such as OLE operate at the user level to provide structures and tools that allow users to actually assemble components.

The definition

of OLE is changing.

Because the migration of OLE services into the COM infrastructure is an evolutionary process, the boundaries between OLE and COM are somewhat ambiguous at any given point in time. However, at the present time OLE is often used to refer to three categories of services that focus on the key elements of component assembly.

❑   OLE automation provides a way for a client application to control another application's objects.

❑   OLE compound documents provide the framework for creating documents containing multiple objects.

❑   OLE controls, also known as OCXs (but now referred to as ActiveX controls), enable components to use OLE automation as a means of exposing properties, methods, and events, which are defined in material that follows.

Two OLE services, automation and controls, are becoming more closely associated with COM. In the process, they are being placed under Microsoft's new category of technologies termed ActiveX. As a result, OLE automation is now sometimes referred to as ActiveX automation, and more frequently as simply "automation." OLE controls are now called ActiveX controls. Eventually, OLE will refer only to services associated with compound documents.

However, there is a bit more involved than a simple renaming of a subset of OLE. The transformation of certain OLE technologies into ActiveX is, according to Microsoft, a "slimming down" of the application framework to improve speed, decrease size, and provide a few additional capabilities.

These modifications are intended to facilitate the use of ActiveX technologies on the Internet, which is becoming an increasingly important platform for the distribution and use of software components. Despite this renaming and slight modification, as the following illustration indicates, the technologies of automation, OLE compound documents, and ActiveX controls still provide the same basic application framework services as their OLE predecessors.

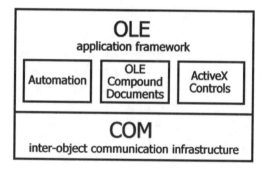

## Automation

Automation refers to the ability of an application to expose objects to another requesting application, and the ability of the requesting application to use the services of the exposed objects. Users of Microsoft Word are familiar with the ability to insert an object, such as an Excel spreadsheet, into a Word document. The spreadsheet object can

---

**Sidenotes (left margin):**

*Portions of OLE are now termed ActiveX.*

*ActiveX includes some modification of OLE.*

*OLE provides tools and structures for assembling objects built on the COM infrastructure.*

*Automation provides access to an application's objects.*

be accessed and edited directly from the Word document. The underlying mechanism used to perform this is automation.

Automation provides for "componentized" applications.

One intent of automation is to support the "componentizing" of previously monolithic applications. Large and comprehensive applications such as spreadsheets and word processors are increasing in size and complexity, to the point where they are delivering more than the typical user can either use or assimilate.

The solution to this functional overkill is to construct applications as collections of objects that can be used independently from the monolithic application. Excel, for example, is constructed in this way. Required objects and functions, such as a spreadsheet and its basic editing functions, are exposed by Excel for use by other applications. The spreadsheet and its editing functions can be accessed from Word when needed, without having to start the entire Excel application.

Automation involves objects, servers, and controllers.

The exposed objects of an application supporting automation, such as an Excel spreadsheet, are referred to as automation objects, which are nothing more than COM objects with appropriate interfaces. The application making the automation objects available is referred to as an automation server, which is simply a COM server. The requesting application—Word in the previous example—is known as the automation controller. The actual process of automation occurs through COM: automation objects publish their interfaces in a type library just as any other COM object does.

## OLE Compound Documents

Compound documents are probably the most familiar feature of the OLE/COM model. In the previous example of automation, the Word document functions as a container for several objects, such as the spreadsheet object

Compound documents require containers.

originating in Excel, or a chart object from a charting application that supports automation. Compound documents allow a seamless presentation of multiple objects as a single document, even if they are obtained from different sources. The relationship between automation and an OLE compound document is shown in the following illustration.

*Automation and an OLE compound document.*

Objects can be linked or embedded.

Compound documents in OLE support either linking or embedding. In both cases, the object appears in the container. The difference is that *linking* simply creates a reference from the object in the container to the original data, whereas *embedding* creates a copy of the data and stores it with the container.

Choosing to link or embed an object will affect how the object can be edited. If, for example, an Excel spreadsheet object in the Word container document is linked, the spreadsheet can be edited only by opening a separate Excel window. However, any changes made to the spreadsheet in Excel will be automatically available in the Word document because the Excel object is linked to the original data.

In-place activation provides for embedded editing.

If the spreadsheet is embedded, another OLE service known as *in-place activation* can be used. This allows the user to double-click on the Excel spreadsheet object within the Word document container to make all of the tools of the Excel editor available within the container environment.

## ActiveX Controls

ActiveX controls were known as OCXs (or OLE controls) until Microsoft recently renamed them. OCXs, the OLE/ COM implementation of software components, super-seded the very successful VBXs (Visual Basic controls). VBXs, however, were closely tied to Visual Basic rather than COM architecture. Therefore, OCXs were created to provide a means of implementing components based on OLE and COM technologies.

ActiveX controls are the key OLE technology for creating components in the OLE/COM model. ActiveX enables software component creation by combining several ele-ments of the OLE/COM component architecture. For example, it builds on two OLE technologies, compound documents and automation, and uses the COM infra-structure.

Bringing together what you have learned about OLE/ COM, ActiveX controls may be defined as *in-process automa-tion servers.* They expose automation objects, and support interaction between the objects and a container. Recall that the concept of automation servers and objects comes from OLE automation, whereas the concept of interaction between a container and a server object comes from OLE compound documents.

## Control Embedding and Containment

The compound document element of ActiveX controls allows them to be embedded within a container, like any other object. For example, an ActiveX control supporting chart creation can be accessed by a word processor docu-ment in order to create and embed a chart object in a text document. The control is not a distinct application that can be started and used to create charts; it is simply a col-lection of chart objects and tools that can be accessed by a container. Adding a chart control to a word processor doc-ument, for instance, would provide all of the charting capabilities of the control to the word processor.

ActiveX controls were previously known as OCXs.

ActiveX controls are components.

ActiveX controls can be embedded in containers.

Programming environments, such as Visual Basic, also function as containers. For example, a chart control may be included in a custom application written in Visual Basic. In this way, multiple controls and components can be assembled to create a full-functioned application.

*Compound document support makes Active X controls easy to use.*

Support for OLE compound documents provides benefits for developers and users. Developers benefit because development environments that support a compound document approach allow applications to be constructed via the assembly of components. Developers can incorporate components into an application as easily as Word users embed Excel spreadsheets into their documents.

Users benefit because the resulting application has a seamless-looking presentation. A form with buttons, charts, and other assorted objects may look like a single document, but the form's assorted objects may actually originate from various controls that have been embedded in the form to create a compound document.

## Control Events and Properties

*Controls recognize events.*

ActiveX controls also introduce the concept of an event to allow a control to communicate back to its container. When an event occurs, such as a mouse click on the control, the control notifies the container of the event, along with any other relevant information, such as the location of the mouse click.

To make use of such events, a container asks the ActiveX control which events are supported, and then subscribes to the event list (called the event set). When an event occurs, the ActiveX control sends a notification of the event to each subscriber. A subscribing container then responds to the event in an appropriate manner, such as running a procedure written by the application developer.

Consider the example of a map control that enables map creation and manipulation. The control could support a

Events allow controls to communicate with containers.

number of events, such as a "mousedown" event triggered (or "fired," in component terminology) when a user clicks on a map created with the control. When the click event occurs, the X and Y coordinates of the mouse click location are passed back to the container (Visual Basic, Delphi, Word, or other) for use.

ActiveX controls also introduce the concept of properties. Properties are attributes of the control, such as the height and width of the control. Most controls provide property pages that can be used in a visual development environment to change control properties. Properties can also be changed programmatically. A chart object from a chart control might have properties such as chart size, chart type (pie, bar, or other), and color. Changing these properties would change the presentation of the object. The following illustration shows an example of an ActiveX control within a container.

Properties are attributes of a control and its objects.

An activeX control within a container.

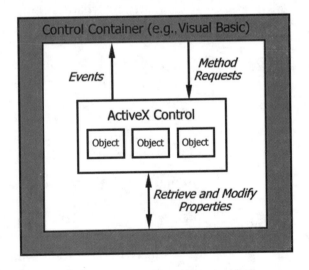

# The OMG and the CORBA Model

**The OMG is a consortium formed in 1989.**

In 1989, the Object Management Group (OMG) was formed with a mission to develop a standard architecture for object-based distributed computing. Trends in distributed computing indicated that component technology was rapidly approaching. In many ways, the group formed in response to these trends indicating the advent of the component world.

**The OMG produced the CORBA model.**

The OMG was formed as a nonprofit consortium of eight member companies, including 3Com Corporation, Hewlett-Packard, and Sun Microsystems. The group has grown to over 700 companies at the time of this writing, and continues to function as a consortium dedicated to developing standard component specifications. Common Object Request Broker Architecture (CORBA) is the result of the group's efforts. CORBA is the OMG version of a distributed object computing model that enables objects and the applications assembled from them to communicate with one another regardless of their location or origin.

## The CORBA ORB, an Object Infrastructure

**CORBA has an object request broker.**

The Object Request Broker (ORB) is central to CORBA specifications. CORBA 1.1 (introduced in 1991) and CORBA 2.0 (adopted in December of 1994) together define the ORB as a sort of middleware that enables client/server relationships between objects. As with COM or any other ORB, the CORBA ORB's purpose is to provide the software bus over which objects can communicate.

**The CORBA ORB provides an infrastructure.**

The ORB works by first accepting a client's request for an object action. The ORB then finds an object that can fulfill the request. The object fulfilling the request may be on the same machine or remotely located over a network. The ORB ensures that the location—as well as the component's language, operating system, hardware, and manufacturer—are transparent to the client requesting an action.

The ORB then passes the parameters of the request to the object it identified as able to fulfill it, invokes the object's appropriate method, and returns the results. The CORBA ORB's internal processes are different from those of Microsoft's COM ORB, but the CORBA ORB's function as an infrastructure for inter-object communication is the same.

## CORBA Object Interfaces

CORBA provides a specification for the publication of object interfaces. CORBA provides an Interface Definition Language (IDL) as a way of defining interfaces. Once again, although the CORBA IDL language is not the same as Microsoft's IDL language, its function as an interface specification language is the same. The CORBA IDL is limited to declarations, and contains no procedural structures or variables. This is in keeping with IDL's purpose, which is to declare interfaces for objects, not implement them.

*CORBA's IDL defines interfaces.*

The CORBA IDL is language and operating system independent, which supports the ORB's independence from language, operating system, hardware, and vendor boundaries. IDL interface definitions are stored in the Interface Repository, the CORBA counterpart to the COM type library. Similar to COM, the CORBA ORB receives a client object request, and then uses the repository to locate an object that can service the request.

*Interfaces are stored in an interface repository.*

## CORBA Common Object Services and Facilities

CORBA also defines Common Object Services (known as *CORBAservices*) as part of the ORB specification to provide system-level services an object can access and make use of. These include such services as licensing and secu-

Common system services are predefined.

Common application services are predefined.

CORBA is a specification.

CORBA ORB, CORBAfacilities, and CORBAservices.

rity. Objects using CORBAservices do not need to implement these services on their own; they can simply access them via the ORB.

CORBA defines Common Facilities (also known as *CORBAfacilities*) that provide generic functions at the application level. Horizontal facilities provide widely used services such as printing and e-mail. Vertical facilities provide services that are specific to particular industries or vertical markets, such as health care and banking. As with CORBAservices, these generic functions can be accessed via the ORB, relieving the client object from implementing them.

The following illustration shows the relationship among the CORBA ORB, CORBAfacilities, and CORBAservices. It is important to remember that CORBA is a specification, not an implementation, of a component model. Implementations are produced by vendors based on the specification of the CORBA model. Over a dozen CORBA-compliant ORBs are currently in existence, including Distributed Objects Everywhere (DOE) by Sun Microsystems, Orbix by Iona, and the System Object Model (SOM) by IBM.

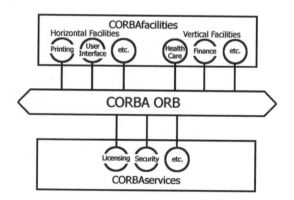

## An Application Framework: OpenDoc

*IBM's SOM is a CORBA implementation.*

SOM, available from IBM as a product called SOMobjects, is closely associated with OpenDoc. OpenDoc was developed by Components Integration Lab (CI Labs)—a consortium including Apple, SunSoft, Oracle, IBM, and Xerox—to provide an application framework for assembling objects into applications. OpenDoc and OLE are conceptually similar because they attempt to achieve very similar goals, but are different in their underlying technical approaches.

*OpenDoc is an example of a CORBA-based application framework.*

OpenDoc uses the compound document approach to assembling components, making it comparable to OLE compound document technology. OpenDoc parts are the components of the OpenDoc world, and so are comparable to ActiveX controls. The Open Scripting Architecture (OSA) of OpenDoc enables parts to expose their objects, and is therefore comparable to OLE automation. Because OpenDoc sits on top of a CORBA-compliant ORB (in this case, SOM), and because it is similar to Microsoft's OLE, the OpenDoc/SOM combination (shown in the following illustration) is a direct competitor to OLE/COM.

*CORBA implementation with OpenDoc and SOM.*

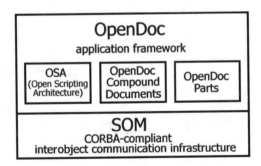

In summary, CORBA provides a significant design for the actual implementation of the component world. It provides the infrastructure and interfaces missing from the

CORBA is an important

component architecture.

traditional object model. It also allows application frameworks to be developed that use an infrastructure and interfaces, ultimately allowing software developers to combine components into working applications.

# Inter-ORB Gateways

CORBA 2.0 supports inter

CORBA ORB communication.

Because CORBA is a specification, not an implementation, it follows that multiple implementations of CORBA, such as SOM from IBM and Orbix from Iona, should be able to communicate easily with one another. What is being done to ensure that diverse ORB implementations can work cooperatively?

CORBA 1.1 focused only on creating specifications for the ORB infrastructure and inter-object communication on a single implementation of that specification (a single ORB). CORBA 2.0, however, focuses on interoperability between multiple implementations of the CORBA ORB. This allows, for instance, objects on a SOM ORB to access services from objects on an Orbix ORB.

IIOP is an inter CORBA

ORB specification.

Inter-ORB object communication is accomplished by the specification for the Internet Inter-ORB Protocol (IIOP). A standard set of message formats have been defined for inter-ORB object communication, essentially supplying an inter-ORB interface language (remember IDL?). The IIOP is set of specifications for how standard message formats can travel between ORBs using TCP/IP.

The specification enables a truly distributed application environment, in which a client-server architecture can be implemented as objects distributed across multiple, networked ORBs. The Internet appears to be the foundation on which many organizations are attempting to implement such applications. (Hence the inclusion of the term *Internet* in the CORBA IIOP specification.)

*Objects can communicate*

*across CORBA and COM.*

If multiple implementations of the same ORB specification are interoperable, what about object communication between ORB implementations based on different specifications? Can objects communicate between a COM ORB and a CORBA ORB, for example?

This is an important issue given the relative technical and market strength of both architectures. Currently, most CORBA ORB implementers provide gateways between COM and CORBA. Unfortunately, these gateways are proprietary to each vendor. Microsoft and the OMG, however, have recently begun work on a specification for a standard COM/CORBA gateway, as shown in the following illustration.

*CORBA 1.1, 2.0, and COM/CORBA interoperability.*

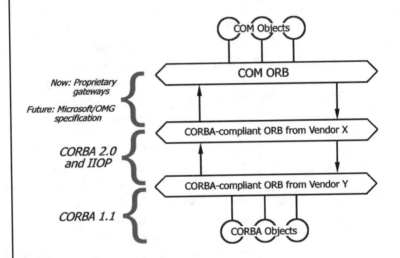

**COM or CORBA?**

This chapter has been intended as an informational discussion of the two leading component models currently available. Greater detail has been provided for the OLE/COM model for the sole reason that the majority of current geographic software component products are based on the OLE/COM model. This market situation is most likely due to the business need of vendors to focus on the most expan-

Current geographic
components focus
on OLE/COM.

sive market possible, which at this point is the Windows
desktop and OLE/COM.

Despite the apparent market strength of OLE/COM in
the domain of geographic components, it should be
stressed that CORBA provides an excellent architecture
for implementing software components, including geographic components. As a result, it is conceivable that the
current market situation could change.

Industry component
specifications should
be ORB independent.

Groups developing component specifications for vertical
domains (such as health care or finance) would do well to
develop specifications that have as little dependency on a
specific component architecture specification (OLE/
COM, CORBA, or other) as possible. This is exactly what
the OGC is doing for the field of GIS.

OLE/COM have different
development approaches.

The OMG approach to developing CORBA focused on
specifying standards prior to product implementations.
This was done to lay a foundation for stable and compatible
implementations. Microsoft, on the other hand, created
the specifications for OLE/COM as the implementation
proceeded (see the section titled "A Brief History of OLE/
COM" earlier in this chapter).

Some see this approach as a case of "putting the cart
before the horse." Others see it as a valid means of testing
specifications against reality as they are being developed,
in a sort of prototyping approach that results in superior
implementation.

Both models will continue
for the foreseeable future.

In the end, a wide variety of technical and market factors
will determine which component model prevails. Important technical factors include how well a component
model integrates with new technologies such as the Internet and object-oriented databases.

Market factors include the release of supporting component development tools, development of industry partnerships, and market education. Regardless of which

component model prevails, it is clear that for the foreseeable future the OLE/COM and CORBA models will both be significant forces in the move toward software components.

# 5

# Geographic Software Components

Because software components are an invention of the software industry in general rather than the GIS industry in particular, it is important to understand the technical foundations of software components from a general technology perspective. In order to provide this understanding, previous chapters have focused on the foundations of software components in general. This chapter focuses on what the GIS industry is doing to incorporate the concepts of software components into its technology offerings.

Software component concepts were developed in the general IT industry.

This chapter reviews the incorporation of software components into the GIS industry at two levels. First, specifications for geographic software components are being produced by groups such as the OGC. These are being developed to provide industry-wide foundations for geographic software components that satisfy the unique needs of the GIS market while adhering to the general IT industry's standards for component architectures. Second, a first generation of geographic software component implementations is being produced by product vendors.

Standards groups, vendors, and users are applying software component concepts in GIS.

## Industry-level Component Standards

Industry-level standards ensure the availability of industry-specific services.

Chapter 4 described how *system-level specifications* for a component architecture provide the infrastructure over which components communicate. You have also seen how *application-level specifications* (application frameworks) such as OLE provide structures and tools with which components based on a given system-level specification, such as COM, can be assembled into applications.

However, these specifications alone will not realize the full value of a component approach. In a given industry or field of interest, *industry-level specifications* must be developed to ensure that components within that industry provide the common set of services and interfaces required by the industry. The differences among the various levels of specifications are represented in the following illustration.

*System-, application-, and industry-level specifications.*

> **Industry-level specifications**
> guidelines for industry-specific
> component services
> (e.g., Health care: HL7, GIS:OGIS)

> **Application-level specifications**
> tools, and structures for component assembly
> (e.g., OLE and OpenDoc)

> **System-level specifications**
> infrastructure for component communication
> (e.g., COM and CORBA)

The leading component architecture designers recognize the importance of industry-level component specifications. The OMG provides for them in its CORBA model through the concept of vertical domain CORBAfacilities (discussed in Chapter 4). Microsoft provides for them in

OMG and Microsoft support industry-level standards.

its OLE/COM model through its OLE Industry Solutions group, which provides active technical and marketing support in the development and progress of industry-specific standards groups.

Components developed according to an industry-level specification can be "certified" by standards-setting organizations as providing the services and interfaces identified as essential to that industry. In addition, the user of a component developed according to an industry specification can be assured that it will integrate with any other component developed to the same industry specification. The result is that those within a given industry can assemble components to create something more than generic applications; they can assemble industry-specific components into applications that meet the very specific needs of the industry.

Industry standards enable the assembly of industry-specific applications.

## An Example of an Industry-level Component Standard

Hypothetical standardization of computer game board and game piece interfaces.

Charlie Kindel, program manager for OLE Industry Solutions at Microsoft, uses the hypothetical example of the computer game market to illustrate industry-level component standards. It may be of interest to the entire computer game industry to define standard interactions between "game boards" and "game pieces." Standard interfaces for common activities, such as the movement of pieces, could be defined and provided as part the computer game industry's industry-level component specification.

Because the "game piece movement service" interface would be standardized, each vendor could develop its games as components using the game industry-standard specification and be assured that its game boards could access the "game piece movement" services of the "game piece" components in other vendors' games.

Industry-level component specifications would support interoperability.

Vendors, users, and the entire industry benefit from a component approach that is fully defined at all levels: system (e.g., COM), application (e.g., OLE), and industry (e.g., a game industry standard). Without such standards, products become closed to interaction with other products in the industry. In the computer game industry example, if vendors develop games with game components developed to the industry's component specification, each vendor's game pieces can interact with other vendors' game boards.

The following illustration depicts how interfaces developed to industry-standard specifications could allow game board and game piece components to interact with one another regardless of the specific implementation of the board and piece movement services. For example, a game board component that is "aware of" the industry specification (vendor A's component in the following illustration) "knows" how to interact with a game piece component that exposes its services through an interface compliant with the specification (vendor B's component in the following illustration).

Standard game industry interfaces could support game/piece interaction.

Game piece components that implement the necessary services but do not expose them according to the specification—such as vendor C's component in the following illustration—become proprietary. They are closed to any other component not specifically designed to work with them.

*An example of component interaction via industry-standard interfaces.*

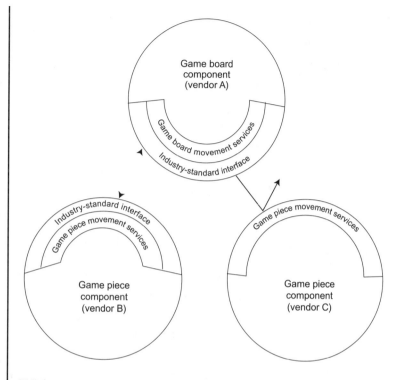

**Game boards and pieces could be mixed and matched.**

With game industry-standard specifications for interfaces between basic components and services common to the game industry, game users could even assemble their own games from boards, pieces, and other components produced by various game vendors. This provides obvious benefits to game users. It also benefits the entire computer game industry by producing new market opportunities.

**Industry standards-setting groups are required.**

To accomplish the development of industry component standards, there must be a standards-setting group within a given vertical market. This group must be broad based enough to incorporate the needs of multiple vendors and users. For example, in the healthcare industry the HL7 (Health Level 7), a widely accepted standards body of healthcare providers and healthcare information system vendors, is working to define standard components for that industry based on OLE/COM.

## GIS Industry Component Standards

This section discusses the Open GIS Consortium, its OpenGIS specification, and application of the specification. Three major parts of the OpenGIS specification are discussed in general terms: the Open Geodata model, which addresses issues regarding the representation of geodata; the OGIS Services model, which addresses geoprocessing; and the Information Communities model, which addresses data sharing.

## The Open GIS Consortium

*The Open GIS Consortium (OGC) is a not-for-profit GIS standards body.*

The OGC was formed in 1994 to fulfill the role of a standards-setting consortium within the GIS industry. The OGC is a not-for-profit consortium that envisions the full integration of geographic information and technology into mainstream computing. The membership of the OGC includes representatives from a wide range of public and private agencies.

Members from the GIS community include such companies as ESRI, Intergraph, and MapInfo. University members range from UCLA to the University of Zurich. Technology industry representatives include Hewlett-Packard, IBM, Oracle, and Sun Microsystems. It is notable that Microsoft and the OMG, proponents of the two dominant component architectures, are also members of the consortium.

*OGC develops interoperability specifications for GIS.*

The OGC recognizes that the infrastructure for mainstream computing is evolving into an interoperable, distributed computing, component-based model defined by technology industry standards such as COM and CORBA. In keeping with this recognition, the activities of the OGC have a dual focus. The OGC is focused on the development of specifications that will enable GIS technologies to be interoperable with one another, which may be viewed as the *internal* integration of the GIS industry.

The OGC is also concerned with the synchronization of these specifications with the interoperable computing specifications of the larger technology industry. This might be viewed as *external* integration with the broader technology context within which GIS exists.

This section discusses the OGC and its work in support of the development of interoperability specifications for geographic technology, with specific attention to issues relating to software components. Much of the information that follows is derived from a series of articles by the OGC appearing in *GIS World* magazine from August 1995 to August 1996; "The Open GIS Approach to Distributed Geoprocessing," an article by Kenn Gardels; and the OGC's *OpenGIS Guide: Introduction to Interoperable Geoprocessing,* edited by Kurt Buehler and Lance McKee.

*The OGC mission includes but is not limited to software components.*

## The OGC's OpenGIS Specification

*OGIS is the interoperability specification produced by the OGC.*

The OGC has developed a specification for a software framework that will support its goals for distributed and open access to geographic software and data. This specification, called the Open Geodata Interoperability Specification (OGIS), provides guidelines that will enable developers of geographic resources (e.g., software vendors and data providers) to create and deploy those resources in a way that ensures their interoperability with other geographic resources. OGIS will eventually be used for certification, allowing developers to market their resources as OGIS compliant.

Chapter 4 discussed the importance of developing industry vertical-market component specifications with as little dependency on a particular component architecture specification as possible. The OGC has worked to create the OGIS specification in this way. The OGC refers to OGIS itself as an "abstract" specification that does not

OGIS is an
"abstract" specification.

depend on any one component architecture, such as COM or CORBA.

OGIS refers to these architectures as distributed computing platforms, or DCPs. It specifies in detail *what* services must be implemented to satisfy the needs of interoperable geoprocessing, but does not specify exactly *how* they must be implemented by each resource provider. (This relationship is shown in the following illustration.) At this level of abstraction, OGIS-compliant software components can be created within any component architecture.

*The relationship between the OGIS abstract specification and implementation specifications.*

| CORBA implementation specification | OLE/COM implementation specification | OTHER implementation specification |
|---|---|---|
| OGIS abstract specification | | |

*OGIS implements in any component architecture (e.g., COM or CORBA).*

At another level, implementation specifications will be developed to define how the OGIS abstract specification can best be implemented within a given component architecture. These are being defined as vendors implement the abstract specification within a given component architecture, using the OGC as a forum for coordination of efforts.

For example, a partial OGIS implementation in CORBA is being created at UCLA. This is being performed within the OGC context and is being observed by the consortium's members. As a result, it is contributing to an implementation specification for CORBA-based, OGIS-compliant geographic software components.

For interaction of geographic resources across component architectures (such as an OLE/COM geographic software component communicating with a CORBA geographic software component), OGIS is relying on the

OGIS components will operate across ORBs.

work of Microsoft, the OMG, and others in the broader technology industry to provide the inter-ORB communication capabilities discussed in the last section of Chapter 4. Developers implementing OGIS-compliant software components in a given environment will rely on system-level inter-ORB capabilities. As discussed in Chapter 4, inter-ORB gateways allow components to work together across multiple-component environments.

The scope of system- and application-level specifications is narrow.

Industry-level specifications are broader.

Because they exist solely to enable interoperable software, system- and application-level specifications for component architectures can focus rather narrowly on the issue of software components, their communication, and their assembly. In contrast, an industry-level specification must include standardization of a broader context.

This broader level of standardization must ask and answer questions such as: What types of data are unique to the industry and how must they be handled? What types of services are specific to the industry? How can resources be shared across groupings of users within the industry?

For example, in the GIS industry, what spatial data types (e.g., lines, polygonal areas, and three-dimensional solids) should be defined? How should geographic reference systems (e.g., map projections) be described? What groups of users exist (civil engineers, ecologists, and so on), and how can their geographic data be described in a way that enables that data to be shared among groups?

OGIS covers geodata, geoservices, and geocommunities.

To address this broadened scope of standardization, the OGC has divided the OGIS specification into three parts: the Open Geodata model, the OGIS Services model, and the Information Communities model (shown in the following illustration). The following sections provide brief descriptions of each of these elements of the OGIS specification.

These descriptions are not intended as a detailed technical guide; rather, they are intended to familiarize you with the concepts and terms of the OGIS specification. Understanding these terms and concepts will become increasingly important as geographic software component providers begin to promote their products with reference to the OGIS specification. For detailed technical descriptions, readers are referred to the *OpenGIS Guide*, edited by Buehler and McKee.

*The OGIS specification consists of three main elements.*

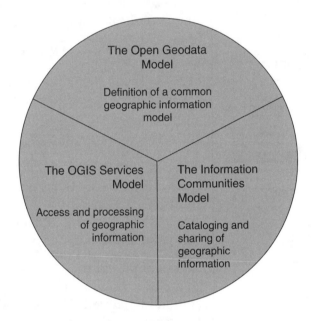

## The Open Geodata Model

The Open Geodata model of the OGIS specification addresses data issues in the GIS industry. The *OpenGIS Guide* describes the Geodata model as defining "a general and common set of basic geographic information types that can be used to model the geodata needs of more specific application domains…". This means that the Geodata model is a conceptual model defining what deodata is and how it might be represented. This is similar to the previous example of the hypothetical computer game industry standards, which would have included a specification for a generalized game piece that could be used by vendors as a model when defining specific game pieces.

The Open Geodata model provides a general structure for geodata.

A unifying geodata model can encompass many specific models.

In defining the Geodata model, OGIS is careful not to restrict *what* can be represented; rather, OGIS provides a model for resource providers to use in defining and creating their own representations of geographic information. This is in keeping with a main objective of the Geodata model, which is to provide a unifying view of geodata. The result is a generalized geographic data model that can encompass many specific data models implemented by data providers and vendors.

The Geodata model defines the basic elements needed to represent geodata.

In the Geodata model, OGIS specifies the basic information types to be used in the representation of geographic data. Geographic occurrences in the real world are viewed by OGIS as *entities* and *phenomena*. *Entities* are recognizable objects such as a well, road, or parcel of land, whereas *phenomena* could be described as geographic events, such as temperature at a given location at a given time. Entities and phenomena have a *location* in time and space, which is represented by *geometry*, such as lines or points. These *features*, which represent real-world objects, can also be aggregated into *feature collections* as appropriate.

The Geodata model recognizes data associated with spatial features.

In addition to being associated with a location, geometry must have a *spatial reference system*, such as a map projection. *Transformations* can be supported to allow conversion between reference systems. Features are recognized as having associated information in the form of *attributes* and *metadata*. Descriptions of information provided about a geographic occurrence—including the geometry, reference system, feature, and metadata—are provided in a *base information definition* specified by OGIS. This provides a commonly understood interface to which users of geographic information can look in order to understand the data and its relevance to their particular applications contexts.

## The OGIS Services Model

The OGIS Services model

addresses geoprocessing.

The OGIS Services model addresses issues related to how services are provided in order to operate on the geographic data defined by the Geodata model. It seeks to define a model for geoprocessing services that provides an environment-independent framework for designing and implementing geoprocessing software.

As such, it directly relates to discussion of software components and the architecture required to support them. Similar to the Geodata model in its domain of geodata, the OGIS Services model provides a specification general enough to provide a unifying approach to the delivery and use of geoprocessing software while still allowing vendors to implement a variety of specific approaches within the overall framework.

Access of and operations

on geodata are both

supported.

Generally speaking, users of geographic information require services to *access* the geographic information, and services to *operate* on it (i.e., analysis and display). Some services, such as transforming data from one geographic projection to another, are so fundamental and common that they can be incorporated into the model in some detail. Other services, such as a particular type of volume analysis, for example, may be more specialized and would be developed outside the model but in accordance with its general specifications for services.

To provide access to data, OGIS incorporates catalog services. A data provider makes data accessible by publishing it in a *catalog*. A catalog includes information regarding the data content and structure. A data user, such as an application developer, uses a catalog to locate required data and understand its structure.

Access is provided

through catalog services.

Transformation services then enable the user to put the data in a form usable in a particular application context. This may include map projection transformation and semantic transformation, a concept discussed in the next section, which describes information communities.

After accessing the data, the user is ready to process the data. The OGIS Services model supports processing of geographic data with *types*, *operations*, and *registries*. Types and operations are comparable to the interfaces and methods of the COM model (see Chapter 4). *Operations* are methods that can be applied to geographic data, and *types* are collections of related operations (just as COM interfaces are collections of methods).

Operations are registered in service registries.

A *type registry* stores information regarding available types, much like the CORBA Interface Repository or the COM Type Library stores interface definitions. An *operation registry* stores information regarding specific operations. When a geographic software component is installed by a user, the component registers its services in the registries to make them available for use.

## The Information Communities Model

The OGIS Information Communities model addresses the issue of data (information) sharing among groups of geographic information users. The model recognizes that within the domain of all users of geographic resources there exist distinct groups who view geographic data in different ways.

The Information Communities model addresses data sharing.

In the *OpenGIS Guide*, editors Buehler and McKee use the example of a civil engineer and an ecologist viewing a road. The civil engineer views the road as a structure with characteristic structural and material requirements, load capacities, and drainage considerations. The ecologist views the road as a structure affecting the surrounding context of plant and animal populations. In Buehler and McKee's words, the civil engineer and the ecologist "might exchange *data* easily because they use the same software, but they won't define highways in the same way, so exchange of *information* will be limited."

Distinct communities describe the same data differently.

In the OGIS view, those groups who share common language, definitions, and representations of geographic

A catalog defines

a community's

description of data.

information are referred to as information communities. OGIS proposes that *catalogs* include a *semantic* element that describes data in terms relevant to a given information community.

In the example of the civil engineer and the ecologist, the civil engineer's data may identify a road with the attributes of name, pavement type, and width. The ecologist's data may identify ecological barriers, which may include what the engineer defines as a road. Regarding attributes, the ecologist may have no use for pavement type but may require knowledge of the width, which may be defined in the ecologist's lexicon as the barrier span.

Semantic translators

translate between

community descriptions.

A catalog facilitates the sharing of geographic data within a community because it uses common languages and definitions defined by the community. Catalogs can be made available to other information communities via *traders*, which are mechanisms that enable information communities to publish their catalogs and discover the catalogs of other information communities.

*Semantic translators* are proposed by OGIS as a means of providing a "mapping" between the language, definitions, and representations of various information communities. A semantic translator would be defined by information communities working in cooperation with one another.

## Applying the OGIS Specification

To demonstrate how the OGIS specification might enhance real-world activity, extend the previous example of the civil engineer and the ecologist. In this hypothetical example, the Engineering division of Anytown, a local municipality, creates and stores data on local roads using the GenericGIS software product. GenericGIS stores the data in specific, "proprietary" structures optimized for GenericGIS. The data content includes attributes of name, pavement type, and width.

Hypothetical example of a real-world application of the OGIS specification.

In order to make the data accessible to the widest possible set of geographic information users, Anytown publishes the data in an OGIS-compliant catalog. The catalog describes the content and structure of the data in a form consistent with the OGIS Geodata model. Any client that understands the OGIS Geodata model and OGIS catalogs can now understand and assess the data's usefulness for its purposes.

OverlayGIS is a CORBA-based geographic software component developed by vendor A to provide specialized complex spatial overlay services. BufferGIS is a COM-based geographic software component developed by vendor B to provide specialized complex spatial buffering services. OverlayGIS and BufferGIS both use algorithms developed by their staffs of geoprocessing specialists, meaning that the software implementations are specific to their respective products.

Fortunately, vendor A and vendor B developed their products in compliance with the OGIS Services model so that, when installed on a computer, the products register their services with the Type and Operation registries specified by OGIS. Any client that understands the OGIS Services model can now easily access the services of both geographic components.

An ecologist from EcoSystems, Inc., is performing impact analysis on environmental systems in Anytown and requires road data for the area. The ecologist makes use of an Internet-based search service that understands OGIS catalogs. The ecologist enters search criteria according to project needs and initiates a search. The available catalogs are searched, and the search service discovers that the Anytown Engineering division's catalog references road data that meets the ecologists needs.

OGIS-compliant query services, the engineering catalog, the ecology catalog, and translators cooperatively defined by the engineering and ecology communities are used to retrieve and transform the data. During this process, the data is transformed from the engineering division's projection to the projection required by the ecologist's project, a semantic translation translates the engineering community's lexicon into the ecology community's lexicon, and any required data structure conversion between vendor-specific formats is performed using the OGIS Geodata model information contained in the catalog.

Semantic translation in this process may include, for example, translating the engineer's term *road* as the ecologist's term *ecological barrier*, and *width* as *barrier span*. The translation process might drop pavement type altogether if the ecologist had no use for it.

The ecologist then uses custom software, written by EcoSystem's technical staff, that incorporates the OverlayGIS and BufferGIS geographic software components into a single user interface. The ecologist uses this custom software to analyze the data according to project needs. The software uses OGIS registries and CORBA/COM inter-ORB services to transparently provide the required spatial functions to the ecologist.

# Examples of GIS Software Component Products

*Software component concepts are becoming a reality in GIS.*

This section describes examples of component-based GIS software. The intent is to provide some examples of how the concepts discussed in previous chapters are actually being implemented as working products. These descriptions focus on products from Environmental Systems Research Institute (ESRI).

Many other vendors are producing geographic components of which the reader should be aware, including such offerings as the GeoMedia product from Intergraph, MapX from MapInfo, and GeoView from Blue Marble Geographics. It should also be noted that the inclusion of product descriptions in this chapter alongside the preceding discussion of OGIS is not intended to indicate compliance by these products with the work of the OGC and the OGIS specification.

Also, note that the focus of this section is on the software element of a component GIS architecture. As the OGIS specification indicates, the data and user elements of an open GIS architecture must not be neglected. GIS vendors are working to develop products that support new object-based data models and information sharing services. However, in keeping with this book's focus on geographic software components, these products are discussed as examples of vendor developments in the implementation of GIS software within a component architecture.

New data models and sharing services are being developed.

## Component Suites and Component Parts

The current approach to geographic software components by vendors and users may be divided into two general categories. First, there are components that exist in suites, or collections. The individual components in GIS component suites are highly specialized. For instance, one component may handle data management, whereas another may handle plotting. Although usable individually, the components are designed to be combined in a complete and comprehensive GIS system.

Component suites are collections of specialized GIS components.

Component suites, as shown in the following illustration, are  at this time generally supplied by large, established vendors in a preassembled form, with a user interface that presents the suite as a coherent application or set of applications. The single-vendor nature of these suites is due in part to the fact that GIS industry-level component standards have not yet matured enough for the inter-vendor interaction of components in a "mixed" component suite. The GIS industry-level standards discussed earlier are in too early a stage to support such interaction.

Component suites are delivered preassembled.

*An example of a GIS
component suite.*

| User Interface (Vendor A) | |
| --- | --- |
| Data management component (vendor A) | Editing component (vendor A) |
| Analysis component (vendor A) | Plotting component (vendor A) |

Second, there are components that are developed and sold as distinct and individual solutions. Instead of being specialized like the individual components found in suites, these component "parts" usually provide a generic set of GIS and mapping services in a single component. They are intended to be used as individual "parts" without regard to a more comprehensive GIS system. For instance, a single component may include capabilities to create maps, display maps in various forms, perform spatial analysis, and plot hard-copy maps.

<u>Component parts are</u>

<u>usually generic in function.</u>

Although component parts, such as the one shown in the following illustration, are currently most often provided by large vendors, they are also provided by smaller vendors, and even by individual developers, who function as "parts shops" to the software industry. Relatively new to the GIS industry, the small vendor "parts shop" has existed in the general software industry for several years, in the form of VBX and OCX component providers.

<u>Component parts are</u>

<u>delivered individually.</u>

*An example of a GIS component part.*

```
┌──────────────────────────────────────────────┐
│        Application Programming Interface       │
│  ┌──────────────────────────────────────────┐ │
│  │ General GIS and Mapping                  │ │
│  │ Component                                │ │
│  │                                          │ │
│  │  • Basic data management services        │ │
│  │  • Basic editing services                │ │
│  │  • Basic analysis services               │ │
│  │  • Basic plotting services               │ │
│  └──────────────────────────────────────────┘ │
└──────────────────────────────────────────────┘
```

*An example of a GIS component part.*

Both of these approaches are excellent news for developers and users who require that GIS software be scaleable, integrated, and resource conservative, as described in Chapter 2. Developers and users with these concerns are the early beneficiaries of the first geographic software components. This is because these needs are met by the *system-level* and *application-level* component standards that are the current focus of GIS component parts and suites.

*Current product focus is on system- and application-level standards.*

Standards at this level allow components to integrate with one another without requiring the specific industry knowledge provided by *industry-level* standards. For example, a spreadsheet component does not need to know anything about the GIS industry to coexist in an application with a geographic software component.

*Standards mean new capabilities for integrating GIS into other applications.*

Many of the early users of GIS software components are integrating GIS capabilities into non-GIS applications. For these users, the distinction between component suites and component parts is not significant because their primary focus is on the individual components and their incorporation into non-GIS software development environments. In this case, it is often of no concern whether a particular component was delivered individually or as part of an assembled collection.

In contrast, traditional GIS users, whose primary tool is the comprehensive GIS, will find the benefits of geographic software components, though not far off, still on the horizon. As the GIS industry develops and implements *industry-level* component standards, such as the OGC's OGIS specification, the "mix-and-match" concept of components *within the GIS software industry itself* will become reality.

Instead of simply cooperating with other non-GIS components, GIS components from multiple vendors will integrate and cooperate with one another (recall the example of the computer game industry described earlier in this chapter). Components that integrate in this way might be termed multipurpose components, various functions of which are represented in the following illustration.

*Multipurpose components will function as individual parts, vendor-delivered suites, or user-assembled mixed suites.*

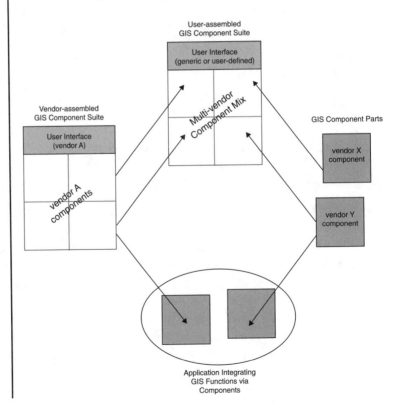

Industry-level standards will provide new capabilities to traditional GIS users.

The flexibility of these interactive components will enable GIS professionals to build comprehensive GIS toolkits. Such toolkits will be built from components that provide the appropriate level of functionality for each functional area, with these combined into a user-assembled suite.

For example, one user may require complex spatial analysis functions and simple map display functions, whereas another may require simple spatial analysis functions and complex map display functions. In the past, both users might have been required to buy the same "one size fits all," single-vendor GIS package. With components, both users will be able to assemble components from various sources into GIS toolkits appropriate to their needs.

Users will create "mix-and-match" GIS toolkits.

## Component Suites

Large GIS systems can be broken into components.

The concept behind component suites is that large GIS toolkits can be broken down into groups of functionality. These sets of functionality can be identified and implemented as components to provide the benefits described in previous chapters. These components can be used individually, as when a non-GIS application requires only one or two spatial functions that can be provided by a single component.

A component suite can also be used in its entirety as a traditional comprehensive GIS system. The engineering of new large-scale GIS software packages and the reengineering of existing ones according to a component approach represents a significant increase in flexibility to users, as previously described. It also represents an opening up of traditionally proprietary GIS software to become accessible through industry-standard APIs and popular software development environments such as Visual Basic and Delphi.

Software components can be used individually or together.

## ESRI's ARC/INFO Open Development Environment

ESRI's ARC/INFO GIS, introduced in 1981, was one of the first commercial GIS products to enter the market. As a pioneering vendor in the GIS software market, ESRI has faced the challenge of evolving its product line through the niche and enterprise eras of the GIS industry, which are described in Chapter 2. ESRI has recognized the trend toward embeddable, scaleable, and resource-conservative GIS as part of the emerging infrastructure era of GIS. In response, ESRI is reengineering its ARC/INFO product from the classic "closed toolkit" (not easily integrated with other products or technologies) to an open component suite.

ESRI is evolving its product line toward components.

Over the years, ESRI's ARC/INFO software has been subdivided into modules oriented toward general categories of tasks. For example, ARCPLOT is a set of map production and plotting tools, ARCEDIT provides tools for maintenance of geodata, ARCGRID provides raster GIS capability, and ARC itself provides geodata management tools.

Major ARC/INFO modules are now accessible as components.

ESRI now provides an implementation of these modules as C/C++ DLLs for Windows and shared libraries on UNIX. The DLLs and shared libraries can be accessed directly. The Windows DLLs, because they support OLE/COM, can be accessed as OCX components in the OLE environment. ESRI has termed this new architecture the ARC/INFO Open Development Environment (ODE).

ARC/INFO components are provided as DLLs and OCXs.

Developers who want to use industry-standard visual development environments that support the OLE/COM component model—such as Visual Basic, PowerBuilder, and Delphi—can use the OCXs as "plug-and-play" components to add GIS functionality to their applications.

Individual components can be used independently.

The developer of a PowerBuilder application, for example, may be required to deliver some advanced map display capabilities in the application. The ARCPLOT component of ARC/INFO could be incorporated into the Power-Builder environment to deliver map display capabilities.

*The new model is called the ARC/INFO Open Development Environment (ODE)*

The currently available ODE components are the ARC, ARCEDIT, ARCPLOT, and ARCGRID modules of ARC/INFO. To access GIS functions from the development environment, command strings are passed to the ARC/INFO ODE components. For example, a command to "build" a particular ARC/INFO coverage can be passed to the ARC component. The ARC component will then perform the requested function using ARC/INFO.

*ODE Objects are Visual Basic classes using ODE.*

To provide an object-oriented framework for using ODE, ESRI has created ODE Objects. These are Visual Basic classes that group related ODE calls as object classes. Therefore, for instance, a "coverage" class provides methods to create and manage ARC/INFO coverages. The command passing to the ODE component is handled by the methods of the ODE Object class. Therefore, rather than directly interacting with the ODE component by passing a string, a developer simply calls methods of the ODE coverage object class to create a coverage object, build it, or perform any other actions supported by the ODE Object class.

ODE Objects are currently supported only in Visual Basic. Similar classes could be implemented by developers in other development environments. The current implementation of ODE and ODE Objects is a stepping-stone to a reengineered, component-based ARC/INFO. As the underlying ARC/INFO code is reengineered to an object model, the objects will be exposed as COM objects with which developers can work directly. Additional ActiveX controls, such as legends and scalebars, will also be part of the developer's toolbox.

*The "plug-and-play" approach anticipates the infrastructure era of GIS.*

There are several points of significance to ESRI's work in this area. First, it represents recognition by a major GIS vendor of the emerging infrastructure era of GIS and its accompanying component suite approach (see Chapter 2). In this environment, GIS functions provided in the form of "plug-and-play" software components will take a major role. Second, it is an example of one approach toward providing a traditional comprehensive GIS toolkit in the new form of a component suite.

New GIS toolkits will take the component suite approach from inception.

The development of traditional GIS toolkits that can be used as comprehensive GISs and/or as individual component parts will likely become an important trend for major GIS vendors and users. ESRI's experience is one of reengineering an existing product to incorporate the functionality of this new approach. In the future, the component suite will be the dominant design on which new GIS toolkits are developed.

## Component Parts

Component parts typically provide fewer capabilities.

ARC/INFO is an example of a comprehensive, end-user-oriented GIS toolkit well suited to the needs of specialists such as GIS professionals. ARC/INFO Open refers to the reengineering of the ARC/INFO product toward a component suite, allowing access to the components by software developers. In contrast to this approach of a typical end-user GIS consisting of a *collection* of components, *individual* GIS components are becoming available to service the needs of software developers whose clients require a limited set of GIS capabilities in their applications.

Component parts are typically for developers.

An important distinction between the component suite and the component part approach is that, unlike component suites, component parts are typically for developers, and are not usually produced in a preassembled form, with an end-user interface. It is left to the developer using the component parts to assemble the appropriate objects from the component, develop an interface, and deliver the finished application to the end user.

## ESRI's MapObjects

MapObjects is a GIS software component based on Microsoft's OLE/COM component model. It is implemented as an ActiveX control providing ActiveX automation objects that deliver GIS functionality. Examples of the automation objects provided by MapObjects are MapLayer, Point, Table, and Address objects. Each of these objects has properties and methods, allowing a developer to manipulate the objects to view and analyze geographic information.

*MapObjects is an ActiveX control with automation objects.*

Because it is based on the OLE/COM model, MapObjects is usable in any programming environment that supports OLE/COM, including Visual Basic, PowerBuilder, Delphi, and Visual C++. Because ActiveX controls require a 32-bit operating system, MapObjects runs on Windows 95 and Windows NT 3.51 or more recent, but not on Windows 3.1 or earlier.

*MapObjects is based on Microsoft's OLE/COM.*

To use MapObjects to develop an application with GIS functionality, a developer first adds the MapObjects ActiveX control to the application in the programming environment of choice. The developer then uses the language of the particular programming environment (e.g., Pascal in Delphi, or C++ in Visual C++) to reference the GIS objects within MapObjects, calling their methods and accessing their properties to perform the required GIS activity. For example, the GeoDataset property of a MapLayer object may be set to reference a specific data source, the Visible property of the MapLayer object set to "true" to allow display of the data, and the MapLayer's SearchByDistance method called to return features in the data source that are within a specified distance of a location.

*Developers use objects, properties, and methods provided by MapObjects.*

In contrast to ARC/INFO and some powerful desktop mapping and GIS software, MapObjects provides a limited, though rich, set of GIS capabilities. Advanced capabilities, such as map projection conversions and raster

MapObjects functionality addresses a broader audience than traditional GIS.

analysis, are assumed to be appropriate to a narrower audience and therefore more suited to a specialized GIS toolkit.

MapObjects is intended to service the GIS needs of a large and general audience of users and developers. As such, it includes general functions such as display of multiple map layers, pan and zoom, query and update of attribute information, thematic rendering, and feature labeling.

MapObjects can coexist with other non-GIS components.

A given application may need to include other components, such as a charting component and a multimedia component, that coexist with the GIS component to provide the total functionality required by the application. GIS functions may be the predominant requirement of the application or they may be incidental. The objects and functions provided in the MapObjects component can be used by the developer and presented to the user only to the degree required by the application.

## Component Suites and Parts, and Their Objects

In the future, component objects will be extendable by developers.

As the delivery of GIS capabilities in component form matures, more access will be provided to the individual objects that comprise GIS component suites and parts. Vendors providing a set of objects in the form of a component suite or part will also provide software developers with low-level access to the individual objects within the set. In the case of MapObjects, which is based on the ActiveX/OLE/COM model, this will mean access to the individual COM objects.

For example, an object representing a map may be provided along with methods to perform standard spatial analyses. With access at the object level, a software developer could add another method to the "map" object in order to implement additional custom-developed spatial analysis.

Components will
increase the flexibility
and accuracy of GIS.

As component suites and parts and their objects continue to evolve in their capabilities and their adherence to industry-level standards, they will greatly increase the flexibility and accessibility of GIS. Many applications that could not use GIS in its previous monolithic form will be geo-enabled using scaleable and resource conservative components. Many people who could not use GIS in its previous complexity will begin to use GIS, perhaps even unknowingly, as it is incorporated into familiar interfaces and applications through the integration made possible by a component approach to GIS.

# 6

# Inside a Geographic Software Component

## A MapObjects Overview

This chapter provides a tour through an actual geographic software component by reviewing the design of ESRI's MapObjects product. The chapter includes an overview of MapOjects, including its architecture, the data sources it supports, and how to access objects from it. It is assumed that you have read or reviewed the previous chapters on component architectures, particularly the terms and concepts associated with OLE/COM and ActiveX.

## What Is MapObjects?

Fundamentally, MapObjects is a geographic software component. Recall that there are many different approaches to implementing software components, with CORBA and the OLE/COM-based ActiveX currently the two most popular. Also recall that the introduction to geographic software components in Chapter 5 included discussion of two types of implementations: component suites and component parts. Bringing together the discussion to this point,

MapObjects is an ActiveX-based GIS component part.

MapObjects can be defined as a geographic "component part" implemented as an ActiveX control.

## MapObjects As an ActiveX Component

MapObjects provides GIS mapping and capabilities.

MapObjects consists of an ActiveX control and a collection of dozens of automation objects that provide GIS and mapping capabilities. Recall from discussions in previous chapters that ActiveX controls provide access to sets of objects. ActiveX controls usually focus on a specific functionality, such as an ActiveX control that specializes in creating charts. The collection of objects provided with MapObjects is focused on GIS and mapping functions.

The ActiveX Map control provides access to GIS objects.

Recall that ActiveX controls are designed to be embedded in an application environment, such as Visual Basic. Once embedded, they provide access to their object collection. For example, using MapObjects, a developer might add a Map control to a Visual Basic form. Then the developer might create MapLayer objects associated with the Map control, load them with spatial data, and access the methods of the MapLayer objects to draw the layers on the Map control, zoom in on them, or search for specific features within them.

MapObjects can be combined with other components.

Because all ActiveX controls are based on the same component specification, multiple components can be combined to assemble working applications. For example, a form may contain a database access control to display database records, a Map control from MapObjects to display the map features associated with the database records, and a charting control to display a chart summarizing a selected set of the database records and mapped features.

Together, these controls would present a single application to a user even though the application was assembled from three components from three different sources, as indicated in the following illustration. By combining MapObjects with other controls in this way, GIS and mapping capabilities can be completely integrated with a wide variety of components to create a wide variety of applications.

*MapObjects can be combined with other components to create applications.*

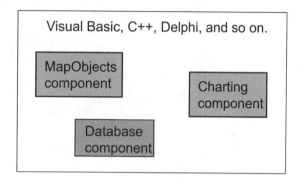

## MapObjects System Requirements

*MapObjects is resource conservative.*

Previous chapters discussed system resource conservation as an important objective of geographic software components. This means that the software itself must not consume inordinate amounts of system resources such as memory and disk space. In keeping with this, MapObjects requires a minimum of system resources. The following are minimum requirements for the software as of the date of this writing.

❏ Processor: Intel 486DX 33 MHz

❏ Operating System: Windows NT 3.51 or higher, Windows 95

❏ Disk Storage: 10 Mb

❏ Memory: 8 Mb

## *The MapObjects Object Model*

The MapObjects design groups its object collection into four general categories that provide the basic functions required for creating, displaying, and working with maps.

❏ *Map display* objects for creating and displaying maps

MapObjects provides four general categories of objects.

❏ *Data access* objects for connecting to and working with data sources used to create maps

❏ *Geometric objects* for drawing and using shapes on maps

❏ *Address matching* objects to return the geographic position of addresses

The following section briefly reviews each of these major MapObjects object groups to help you understand what MapObjects provides. Included on the accompanying CD is a detailed diagram of the MapObjects object model that you may view and print if you would like more detail. The descriptions and high-level object model diagrams that follow are from ESRI's on-line help provided with MapObjects, and from the *Building Applications with MapObjects* guide published by ESRI as part of the MapObjects documentation.

When viewing the diagrams, note that the arrows connecting objects show the major associations between objects within categories. For example, an arrow drawn from the box representing the Map control to the box representing the Layers collection indicates that a Map control may have a Layers collection. Similarly, an arrow from the Layers collection to the MapLayer object means that a Layers collection may have a MapLayer object.

# Map Display Objects

Map display objects are used to create and display maps.

Map display objects allow you to create, display, and manipulate maps, raster images, and dynamically updated maps. The objects are shown in the following illustration.

*Map display objects.*

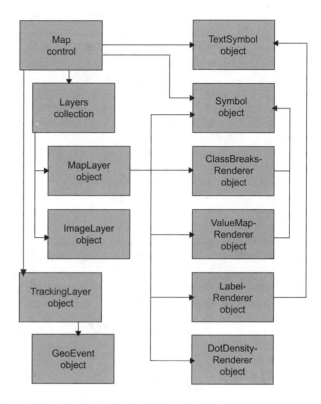

## The Map Control

*The Map control*

*responds to events.*

Any software component must provide a means for the user to interact with the component. In the OLE/COM model, the ActiveX control serves this purpose. The Map control is unique in the MapObjects object collection because it is the ActiveX control of MapObjects. Recall from Chapter 4 that ActiveX controls support *events* to provide a way for an application to respond to a user.

Examples of events may include the user pointing and clicking with a mouse, or pressing a key on a keyboard. An ActiveX control defines events it will recognize, and when one of these events occurs, such as a user clicking on the control, the control responds.

The developer determines

what happens when an

event occurs.

The developer who added the control to the application is responsible for indicating to the control how it should respond to particular events, such as executing a program in response to a user mouse click. This is usually done by adding program code to a procedure provided by the control for a specific event. For example, a MouseDown procedure may be provided for the MouseDown event of a control. The MouseDown event occurs when a mouse key is depressed with the mouse pointer over the control. The procedure, and any code placed in it, will execute when the event occurs.

Events pass information to

their procedures.

When an event occurs, the control may capture information regarding the event and pass it to the procedure. For example, the MouseDown event procedure may have the following declaration:

```
MouseDown(button Integer, shift Integer, x Single, y Single)
```

This means that when the MouseDown event procedure runs in response to a MouseDown event, the variables *button*, *shift*, *x*, and *y* are loaded with values that can be used in the procedure. *Button* will contain an integer indicating the number of the mouse button that was depressed; *shift* will contain an integer indicating whether a combination of the shift, alt, and control keys were pressed while the mouse key was pressed; and *x* and *y* will contain single precision numbers indicating the X and Y coordinate values of the location of the mouse pointer when the mouse key was depressed. The relationship among user actions, events, and programmed procedures is shown in the following illustration.

*How an event links
user actions to
programmed
responses.*

Map control events are used

to make maps interactive.

The events recognized by the Map control enable a developer to create maps that are highly interactive. The user interacts with the map—using the mouse, for example–which triggers the appropriate event, which runs an event procedure in which the developer updates the map according to the user's direction. The following events are recognized by the MapObjects Map control.

**AfterLayerDraw (ByVal As** *index* **Integer, ByVal** *canceled* **As Boolean, ByVal** *hDC* **As OLE_HANDLE)**

Occurs once after each MapLayer is drawn. *Index* contains a value indicating what layer of the map was drawn, *canceled* contains a value indicating whether drawing was canceled by the user, and *hDC* contains a reference to the Microsoft Windows device context of the map.

**AfterTrackingLayerDraw (ByVal** *hDC* **As OLE_ HANDLE)**

Occurs after the TrackingLayer is drawn. *hDC* contains a reference to the Microsoft Windows device context of the map.

**BeforeLayerDraw (ByVal** *index* **As Integer, ByVal** *hDC* **As OLE_HANDLE)**

Occurs once before each layer is drawn. *Index* contains a value indicating what layer of the map is about to be drawn, and *hDC* contains a reference to the Microsoft Windows device context of the map.

**BeforeTrackingLayerDraw (ByVal** *hDC* **As OLE_ HANDLE)**

Occurs before the TrackingLayer is drawn. *hDC* contains a reference to the Microsoft Windows device context of the map.

**Click ( )**

Occurs when a mouse click is performed on the Map control.

**DblClick ( )**

Occurs when a double mouse click is performed on the Map control.

**DragDrop (*source* As Control, *x* As Single, *y* As Single)**

Occurs when another control has been dragged and dropped onto the Map control. *Source* identifies the control that was dropped, and *x* and *y* contain values indicating the position of the mouse pointer when the drag and drop was performed.

**DragFiles (ByVal *fileNames* As Object, ByVal *x* As Single, ByVal *y* As Single, ByVal *state* As Integer, *dropValid* As Boolean)**

Occurs when a set of files is in the process of being dragged and dropped. *FileNames* contains the names of the files being dragged; *x* and *y* indicate the current position of the mouse pointer within the control; *state* indicates whether the files being dragged are entering, leaving, or currently over the control; and *dropValid* indicates whether the Map control on which the files are being dragged is valid.

**DragOver (*Source* As Control, *x* As Single, *y* As Single, *state* As Integer)**

Occurs as a drag-and-drop operation is in progress. *Source* identifies the control that is being dragged, *x* and *y* contain values indicating the current position of the mouse pointer when the drag is being performed, and *state* indicates whether the object being dragged is entering, leaving, or currently over the control.

**DrawError (ByVal *index* As Integer)**

Occurs when a map cannot draw a MapLayer. *Index* identifies the layer that cannot be drawn.

**DrawingCanceled ( )**

Occurs when a map draw operation is canceled by the user.

**DropFiles (ByVal** *fileNames* **As Object, ByVal** *x* **As Single, ByVal** *y* **As Single)**

Occurs when a file drag-and-drop operation completes. *FileNames* contains the names of the files dragged, and *x* and *y* indicate the position of the mouse pointer within the control when the drag completes (when the mouse is released).

**GotFocus ( )**

Occurs when the Map control receives the focus (as by tabbing to or clicking on the control).

**LostFocus ( )**

Occurs when the control loses the focus (as another object receives the focus).

**MouseDown** (*button* **As Integer**, *shift* **As Integer**, *x* **As Single**, *y* **As Single)**

Occurs when the mouse is pressed while the mouse pointer is over the Map control. *Button* indicates the number of the mouse button that was depressed; *shift* indicates whether a combination of the shift, alt, and control keys were pressed when the mouse key was depressed; and *x* and *y* contain single precision numbers indicating the location of the mouse pointer when the mouse key was depressed.

**MouseMove** (*button* **As Integer**, *shift* **As Integer**, *x* **As Single**, *y* **As Single)**

Occurs when the mouse pointer is moved over the Map control. The parameter definitions are the same as for the MouseDown event.

**MouseUp** (*button* **As Integer**, *shift* **As Integer**, *x* **As Single**, *y* **As Single**)

Occurs when the mouse is released while the mouse pointer is over the Map control. The parameter definitions are the same as for the MouseDown event.

## Other Map Display Objects

❏ The Layers collection includes MapLayer and/or ImageLayer objects for each layer of information to be displayed on a map. For example, roads may be one layer, lakes another. Both will exist as members of the Layer collection for the map with which they are associated.

❏ A **MapLayer** object represents a georeferenced data layer on a map that is drawn with features from a GeoData set.

❏ An **ImageLayer** represents a layer in a map based on georeferenced raster data stored in a file on disk.

❏ A **TrackingLayer** object represents a layer in a Map that depicts geographically referenced phenomena whose positions may change.

❏ A **GeoEvent** object represents a geographically referenced phenomenon whose position may change.

❏ A **Symbol** object consists of attributes that control how to represent features or graphic shapes.

❏ A **TextSymbol** object consists of attributes that control how to render text.

❏ A **ClassBreaksRenderer** is an object that represents a way of classifying features into categories or classes, by drawing different symbols for each category established for its Field property.

❏ A **ValueMapRenderer** is an object that represents a way of symbolizing features of a MapLayer by drawing a Symbol for each unique data value.

❑ A **DotDensityRenderer** is an object that represents a way of symbolizing features by drawing dots on a feature to indicate density.

❑ A **LabelRenderer** is an object that represents a way of symbolizing features by drawing text on a feature.

## Data Access Objects

*Data access objects are for accessing source data for maps.*

Data access objects, shown in the following illustration, allow you to establish connections to geographic data and associate it with the layers of a map. With data access objects you can also view and modify the attribute tables associated with geographic data.

*The data access objects.*

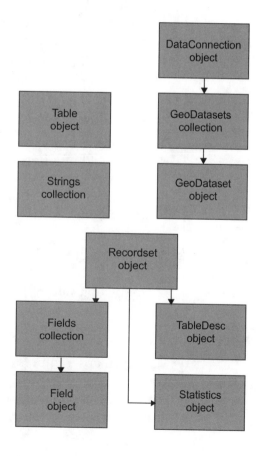

❏  A **DataConnection** represents a connection to a database.

❏  A **GeoDataset** object represents geographic data that may be associated with a layer on a map.

❏  A **GeoDataset** collection contains all GeoDataset objects of a DataConnection.

❏  A **Recordset** object represents the records associated with a GeoDataset or the records that result from running a query.

❏  A **TableDesc** is an object that represents a description of the Fields of a Recordset.

❏  A **Table** object is a read-only data access object. Generally, a Table object is a logical representation of a physical table in a database that contains information about a particular subject.

❏  A **Field** object represents a column of data with a common data type and a common set of properties.

❏  A **Fields** collection contains all stored Field objects of a Recordset object.

❏  A **Statistics** object represents the result of a calculation on a Field of a Recordset using the Recordset object's CalculateStatistics method. A Statistics object has the following statistical properties: Max (maximum), Min (minimum), Mean, StdDev (standard deviation), and Sum.

❏  **Strings** is a standard collection that includes a set of unique String data types. This is frequently used for extracting values from other objects.

## Geometric Objects

Making maps requires the use of geometry. The geographic features themselves are represented by geometric shapes, such as points, lines, and polygons. You might also draw your own geometric shape (such as a rectangle)

Geometric objects are
for representing
geometric shapes.

around an area to designate that particular area as being of interest.

Geometric shapes are also used in GIS to perform selections of features located within a specific geometric shape, such as all restaurants within a three-mile radius of a given address. The radius would be represented by a geometric shape (a circle) centered on the address. MapObjects provides the geometric objects shown in the following illustration for purposes such as these.

*The geometric objects.*

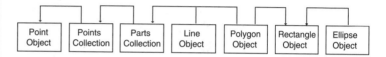

- ❏ A **Rectangle** object represents a geometric shape with four edges and four right angles. It is often used to reference the bounding rectangle of a map's geographic extent.

- ❏ A **Point** object represents a geometric shape that has one point with X and Y coordinates.

- ❏ A **Points** collection contains Point objects for a Line or Polygon object.

- ❏ A **Line** object represents a geometric shape that has two or more points.

- ❏ A **Polygon** object represents a multipart set of shapes or a single-part geometric shape, each part having three or more points.

- ❏ A **Parts** collection represents the set of points representing an exterior or interior boundary of a Polygon or the set of points that make up the sections of a multipart Line. For example, a "doughnut" polygon has two parts: one that represents its outer boundary, and another that represents the "hole." Most polygons or lines are single-part shapes; however, multipart shapes are useful for representing multipart features as single entities, such as the islands of Hawaii or a lake and its island as a "doughnut" polygon.

## Address Matching Objects

A standard activity when working with maps is determining the location of an address. This is required in many of the everyday uses of maps, such as determining how to get to a destination. The activity of taking a supplied address, matching it to an address or range of addresses located in a database, and then displaying the location of the address is termed address matching.

Address matching was one of the first functions provided by early GIS software. MapObjects provides the address matching objects shown in the following illustration to perform address matching.

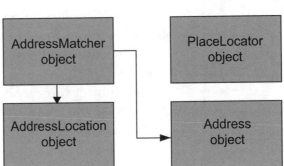

*The address matching objects.*

- ❏ An **AddressMatcher** object lets you specify an individual address or street intersection, or supply a table of addresses, to match against a street network. The AddressMatcher provides methods to perform address matching for these cases.

- ❏ An **Address** object represents a standardized address. MapObjects returns a standardized address with the AddressMatcher object's StandardizeAddress method.

- ❏ A **PlaceLocator** object lets you match place names to a GeoDataset containing a table of place names.

- ❏ An **AddressLocation** object represents the results of an address match.

# Data Sources Supported by MapObjects

Any geographic component must recognize that geographic data comes in a variety of forms. When dealing with geographic data, a user is typically confronted with three fundamental characteristics of information: *representation*, *storage*, and *format*.

## Data Representation, Storage, and Format

Generally, geographic data sources are *represented* in two ways: vector and raster. (See the following illustration.) A vector representation organizes geographic information using the Cartesian coordinate system. This means that locational data is stored as XY coordinate pairs. Points are stored as single pairs of XY coordinates, whereas lines and polygons are stored as ordered pairs of coordinates. Vector representations are often used for representing data with exactly known locations, such as streets, light poles, or the legal boundaries of lots.

*Vector data represents geography using coordinates.*

A raster representation organizes geographic data using cells arranged in rows and columns. Each cell has a row number and a column number, with the cell in the upper left identified as row 1, column 1. Values associated with each cell describe the geographic attributes in the region of space covered by the cell.

*Raster data represents geography using grid cells.*

Raster representations are often used for geographic data with less discrete locational boundaries. This is often the case with environmental data, such as soil type polygons or forest boundaries, where the mapped features may not have sharply definable boundaries.

*Raster and vector representations of geographic data.*

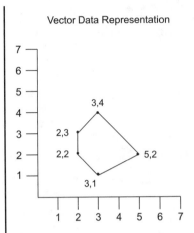
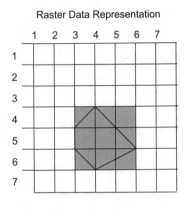

Digital geographic data, whether represented as vector or raster, is usually *stored* in one of two forms: disk files or database tables. Disk files are typically used for small- to medium-size mapping data sets. Very large mapping data sets often require the data management capabilities of database management systems, and may be stored as database tables.

*Geographic data may be stored in files or tables.*

In addition to representation and storage, the user of geographic data must also be concerned with *format*. For example, Windows bitmap, Sun raster file, and Tag Image File are all different disk file formats for data represented in a raster form. MapObjects supports a variety of data representations, storage mechanisms, and formats. These are discussed in the following sections.

*Geographic data can be in various formats.*

# Shapefiles

Shapefiles are a format created by ESRI for storing vector data. Shapefiles are nontopological, which means that limited information is kept in the shapefile regarding the relationships of features to one another, such as what specific line shapes are used to define a specific polygon shape. The lack of topology means that shapefiles are less appropriate for sophisticated spatial analysis than other

*Shapefiles store vector data nontopologically.*

formats. However, it also provides some advantages, such as improved drawing times.

**Shapefiles consist of three disk files.**

The shapefile specification is openly published; therefore, shapefiles can be created by anyone. The shapefile format consists of three elements: *shape, index,* and *attribute.* Each of these elements is stored as a separate file on disk; therefore, a shapefile actually consists of three disk files (one for each element).

The *shape* element is the portion of the shapefile describing the geometric shapes that represent the geographic features. These shapes are described by their X and Y coordinate locations. For example, a line representing a road may be described as a series of X and Y coordinate points: one point for the start of the line, one point for the end of the line, and as many points in between as are required to define the line's shape. The disk file containing the shape element has a file name extension of *.shp.*

**A shapefile contains geometric shapes.**

**An index file indexes geometric shapes.**

The *index* element of the shapefile provides an optimized means of accessing the geometric shapes described in the shape element. The element contains a sequential index of offsets into the shape data. This spatial indexing, as it is called, provides for faster drawing times and faster queries of geographic features represented by geometric shapes. The disk file containing the index element has a file name extension of *.shx.*

**The attribute file contains associated table data.**

The *attribute* element contains tabular data associated with geographic features. For example, a road may have associated information regarding its condition, the year it was constructed, and its width. This information can be stored in the attribute element of the shapefile and associated by a key value with the specific road shape to which it pertains. The attribute element of the shapefile is stored as a standard dBASE file with one record per shape. The disk file containing the attribute element has a file name extension of *.dbf.* The elements of a shapefile are shown in the following illustration.

*The elements
of a shapefile.*

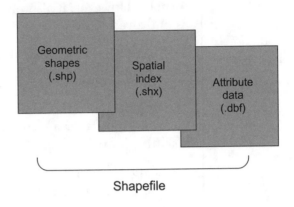

Shapefile

## ARC/INFO Coverages

ARC/INFO coverages store vector data topologically.

ARC/INFO coverages are another format created by ESRI for storing vector data. There are several differences between coverages and shapefiles. First, in contrast to shapefiles, coverages are proprietary data structures. This means that the specification for them is not openly published.

Second, also in contrast to shapefiles, coverages are a topological data structure. This means that the format is much more sophisticated in its ability to track the relationships between features, such as what specific line shapes are used to define a specific polygon shape. Because of their more sophisticated data structure and the inclusion of topological information, coverages are better suited for larger data sets and for applications requiring complex spatial analysis.

Geometric shapes are stored in files in a coverage subdirectory.

ARC/INFO coverages are represented by subdirectories existing within an ARC/INFO workspace. The workspace is a directory that serves as a work area and storage area for the ARC/INFO coverages. The coverage subdirectories each represent a single coverage and are named with the coverage names. These subdirectories contain the geographic data stored in file names such as TIC, BND, and

ARC. This data can be created and maintained using the ARC portion of ARC/INFO.

A separate subdirectory within the workspace contains the attribute data associated with the geographic features of all coverages stored in that workspace. This single subdirectory, named *info*, stores the attributes in database tables that can be accessed using the INFO portion of ARC/ INFO. The following illustration shows an example of an ARC/INFO storage structure.

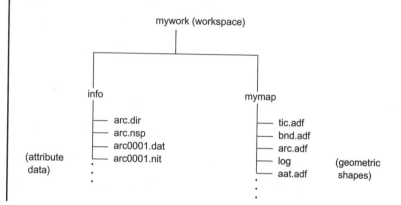

## Spatial Database Engine

The Spatial Database Engine (SDE) is a product of ESRI. SDE is designed to provide high-performance of operations on large spatial data sets. This is accomplished by storing the spatial data within a relational database management system (RDBMS) and providing a set of software services for accessing the data. In this format, spatial data and attribute data are both stored in database tables within an RDBMS. The RDBMS may be any RDBMS supported by SDE, such as Oracle by Oracle Corporation.

Because SDE exists within an RDBMS environment, it leverages the capabilities of the RDBMS to manage large amounts of data. This architecture also brings to the spa-

SDE leverages the strengths of the RDBMS.

tial data the security and data integrity mechanisms that are part of the RDBMS environment. SDE employs a client/server architecture, and provides a C-language API to the software services provided for accessing the spatial data.

SDE data sets are high-level groupings of spatial data.

The data structure of SDE groups data into data sets, layers, and features. SDE data sets group related geographic, tabular, and data-set-specific security information. For example, all geographic data for a given city might be stored in a single data set. A separate DBMS account is created for each data set.

SDE data sets consist of SDE layers. Layers are sets of thematically related geographic data that are subsets of the data set. If the data set were the set of all geographic data for a given city, roads might be one layer, lots another, and so on. SDE layers are "tile-less." This means that the geographic data is not partitioned into "tiles" or pieces that must be reassembled when necessary. Tiling is often done in other storage mechanisms in order to manage large amounts of data. In contrast, SDE stores the data in layers that are logically continuous.

Layers are "tile-less" sets of related spatial data within data sets.

Features are the individual geographic elements within layers; a single road in the roads layer, for example. Just as SDE layers are continuous, SDE features are also continuous. Each road, for instance, is stored as a completed and unbroken feature. Tile-based systems often split features that cross tile boundaries. Because SDE stores features in their entirety, performance is optimized because each feature can be retrieved by a single disk access. The SDE architecture is shown in the following illustration.

Features are individual geographic elements within layers.

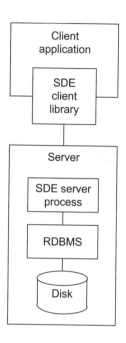

*The Spatial Database Engine architecture (adapted from ESRI SDE product documentation).*

## Image Files

Image files are raster representations of geographic data. They are often used to display aerial photographs and satellite images. If a raster image is to be displayed in real-world units such as feet, the location of the rows, columns, and cells of the cell-based raster image must be mapped to the XY coordinate plane of a real-world mapping coordinate system. This allows image data, such as an aerial photo, to be displayed as a backdrop to vector data, with the vector and raster data properly aligning. This mapping of cells to a coordinate system is referred to as georeferencing.

*Image files represent geographic data in raster form.*

Georeferencing information is usually stored either in the header file of the raster image or in a separate file, known as a "world" file. The information is in the form of values to be used in a standard transformation equation. Various products can be used to georeference an image that is not already georeferenced.

*Image files can be georeferenced.*

MapObjects supports a wide variety of image formats. The following table of supported image formats is taken from the *Building Applications with MapObjects* guide published by ESRI as part of the MapObjects documentation. The illustration that follows the table shows an example of an image file.

| Short Name | Descriptive Name | Image File Extensions | World FileExtensions |
|---|---|---|---|
| BMP | Windows Bitmap | *.bmp | *.bmpw or *.bpw |
| TIFF | Tag image file | *.tiff, *.tif, *.tff | *.tfw |
| SUN | Sun raster file | *.sun, *.ras | *.snw, *.rsw |
| ERDAS | ERDAS GIS or LAN | *.gis, *.lan | *.gsw, *.lnw |
| IMPELL | IMPELL bitmap | *.rlc | *.rlw |
| BIL | Band interleaved by line | *.bil | *.blw |
| BIP | Band interleaved by pixel | *.bip | *.bpw |
| BSQ | Band sequential | *.bsq | *.bqw |

*An example of an image file of a satellite image taken over the state of Washington.*

# Database Tables

*Database tables can store information related to geographic features.*

Locational geographic data usually has related attribute data. For instance, a line representing a road (the locational geographic data) may have associated information regarding its width and pavement material type. Some of this information may be stored in tables directly associated with the geometric shapes, as in the case of an ARC/INFO coverage's INFO files or a shapefile's *.dbf* files. However, in many cases the data may be stored in distinctly separate database tables that must be associated with the geographic data source.

*MapObjects can join database tables to map layers.*

MapObjects provides methods for establishing "relates," which are similar to "joins" in an RDBMS. By specifying a table to relate to a MapObjects map layer—along with key—table-based data can be associated with a map. The Data Connection objects of MapObjects can then be used to work with the new "related" table data. The following illustration shows the joining of database tables with geographic data.

*Joining database tables.*

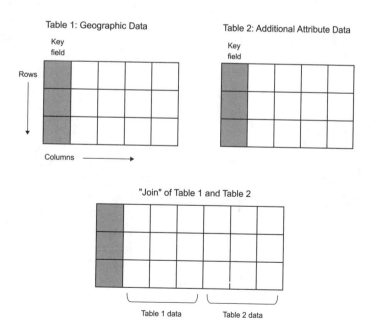

## Data Source Abstraction

Physical references to data sources are encapsulated in objects.

One of the great advantages of geographic components and their object-based approach is that references to data sources can be abstracted. This means that the only physical references in an application to specific disk files and data sets are isolated to the objects that establish connections with data sources. The rest of the application can view the data logically by referencing the data-connection objects that encapsulate the physical references to the data.

Objects can be easily redirected to new data sources.

Changing data source types, such as changing from disk files to database tables, can be done with relative ease to the extent that the component supports various data source types. For instance, MapObjects generally supports the same operations for shapefiles (stored as disk files) and SDE layers (stored as tables).

Only a limited portion of an application needs to know the specifics of the data storage mechanism, because MapObjects encapsulates this in its data connection objects. If an application were built to access shapefiles and those shapefiles were later moved into SDE, the application could remain unchanged apart from some minor changes to the properties of the data connection objects.

## Accessing Objects in the MapObjects Collection

The manner in which you access the objects contained in a geographic component product will vary depending on the implementation of the product (e.g., CORBA or OLE), the development environment you use (e.g., Visual Basic or C++), and your application design (e.g., displaying a map versus simply using the component's capabilities to return values). The following section takes a brief look at how objects within a geographic software component are accessed, using MapObjects and Visual Basic as an example.

## Adding MapObjects to a Development Environment

There are two ways to use MapObjects in Visual Basic.

There are two ways in which MapObjects may be added to a Visual Basic project: as a control or as a referenced object library. Before your specific development environment can access MapObjects in either of these ways, however, MapObjects must be installed on your computer. See the appendix for instructions on how to install MapObjects from the accompanying CD-ROM.

Adding MapObjects as a custom control.

Adding a component as a control will allow you to use ActiveX controls supplied by the component. After adding MapObjects as a control, for example, you will be able to add the Map control to a Visual Basic form. This will provide a place for maps to be drawn and displayed.

Adding MapObjects as an object library.

In contrast, adding MapObjects as a referenced object library will provide you access to the collection of ActiveX automation objects supplied by MapObjects, but without the ActiveX control interface. You would do this if you wanted to access the functionality of MapObjects without needing to draw a map of the results.

For instance, you may want to know which political district a specific address is in. This can be done without drawing a map simply by using the address matching and spatial analysis functions of MapObjects to perform the analysis, then returning the name or number of the political district to the requesting object. In this case, a map control is not required because no map is displayed.

### Adding MapObjects as a Control

Adding MapObjects as a control is done just as with any other ActiveX or OCX. The following steps outline the procedure.

**1.** Open a Visual Basic project. In your toolbox you will see the collection of controls your project is currently enabled to work with. This is the collection to which you will add MapObjects.

2. On the Visual Basic menu bar, open the Project menu, and within it select the *Components...* command. This will present you with a dialog showing several tabs. View the Control tab to see the controls accessible by your project. Each control has a check box to the left of its description. The controls your project is enabled to work with have a check mark in their corresponding checkbox. The Project menu and the *Components...* command are shown in the following illustration.

*The Visual Basic Project menu and the Components... command.*

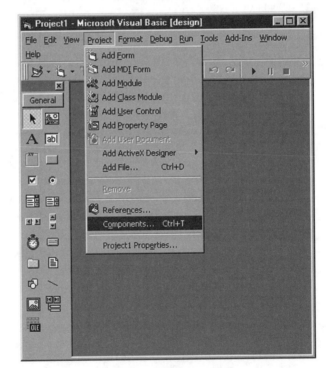

3. Click on the check box next to "ESRI MapObjects" to enable your project to access MapObjects. If there is no "ESRI MapObjects" listed, review your installation process to make sure it was successful.

**4.** Click "OK" or "Apply" in the Components dialog. Visual Basic will add the MapObjects icon to your toolbox, indicating that the MapObjects control and its associated objects are now accessible to your project. The Components dialog and the control reference are shown in the following illustration.

*The Components dialog and the ESRI MapObjects control reference.*

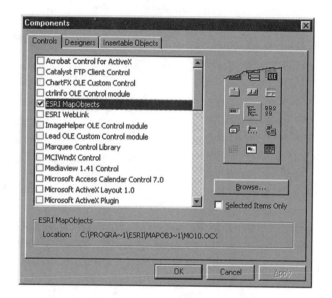

## Adding MapObjects as a Referenced Object Library

Adding MapObjects as a referenced object library is very similar to adding MapObjects as a custom control. The following steps outline the procedure.

**1.** Open a Visual Basic project.

**2.** Open the Project menu, and within it select the *References...* command, located just above the *Components...* command. This will present you with a dialog showing the object libraries currently accessible by your project. Each reference has a check box to the left of its description. The references your project is enabled to work with have a check mark in their corresponding check-box. The *References* dialog and the MapObjects reference are shown in the following illustration.

*The References dialog and the ESRI MapObjects reference.*

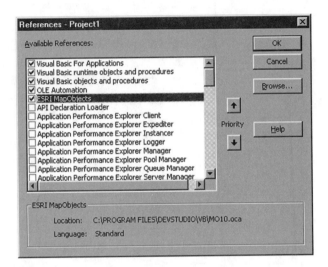

**3.** Click on the check box next to "ESRI MapObjects" to enable your project to gain access to the objects provided by MapObjects. If there is no "ESRI MapObjects" listed, review your installation process to make sure it was successful.

**4.** Click "OK" in the *References* dialog. Your project is now enabled to work with the collection of automation objects provided by MapObjects.

## Viewing MapObjects Object Classes

*A component's objects, methods, and properties can be browsed.*

The help files provided by MapObjects contain complete listings and descriptions of MapObjects objects, methods, and properties. Most development environments provide a means of quickly browsing components, and Visual Basic is no exception. To use Visual Basic's Object Browser for viewing MapObjects object definitions, perform the following steps.

**1.** Open the View menu on the Visual Basic menu bar, then select the *Object Browser* command. The View menu and the *Object Browser* command are shown in the following illustration.

*The View menu and the Object Browser command.*

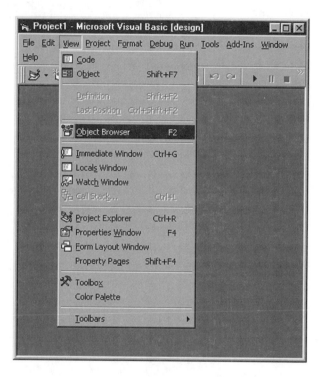

**2.** In the libraries drop-down list, select MapObjects. (See the following illustration.) If MapObjects does not appear, it is likely that you have not added the MapObjects component to the Visual Basic project. Refer to the previous section for more information on this. The *Classes* scrolling list box and the *Members of...* list box will now display information regarding the MapObjects object collection.

*Selecting the MapObjects library in the Object Browser.*

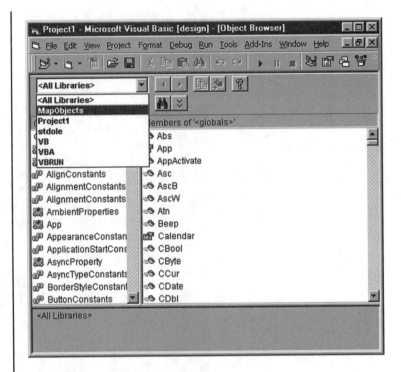

**3.** Select an object class from the *Classes* list box to view the class members, as shown in the following illustration. For example, if you select the "MapLayer" object class, the *Members of...* list box will change its title to *Members of 'MapLayer'* and will be updated to display the methods and properties associated with the object class. Also note that a brief description of the object class appears in a line of text at the bottom of the window.

*MapObjects classes and members in the Object Browser.*

**4.** Right mouse click on the object class, then select the *Help* command from the resulting drop-down menu to open MapObjects on-line help files for that object class.

↦ **NOTE:** *You may experience problems accessing MapObjects help in this way if you have not installed the new MapObjects VB5 help files. These are available on the accompanying CD-ROM.*

**5.** Return to the Object Browser window and select a method or property from the *Members of...* list box. Note that a brief description of the method or property appears in a line of text at the bottom of the window.

**6.** Right mouse click on the method or property, then select the *Help* command from the resulting drop-down menu to open MapObjects on-line help files for that method or property.

Recall from previous chapters that a class is essentially an object definition. Using a class, you can create objects that inherit the characteristics of the class. With MapObjects, classes are already provided, as you have seen by using the Visual Basic Object Browser. These object classes are used to create objects as defined by the classes.

For example, MapObjects supplies a MapLayer object class. Using this class as a template, you can create a MapLayer object to represent a specific layer of data; roads, for instance. The MapLayer object you create would be an *instance* of the MapLayer object class. Review Chapter 2 if you are unsure about the distinction between objects and classes.

Notice that what have previously been referred to as the "objects" supplied by MapObjects (MapLayer, Symbol, and so on) are referred to by the Visual Basic Object Browser as "classes." MapObjects provides you with classes from which you can create objects to use in your application. Therefore, from a developer's perspective, it is not quite technically correct to refer to MapObjects object classes as objects.

The MapObjects documentation refers to the classes as objects because of the terminology of OLE, which describes MapObjects as a collection of *automation objects*. See Chapter 4 for more discussion of the concepts of automation servers and automation objects. To prevent confusion when referencing the ESRI documentation together with this book, MapObjects object classes are referred to as the *objects* provided by MapObjects, unless context indicates otherwise.

---

*Margin notes:*

MapObjects provides map-related object classes.

Classes are used to create mapping objects.

Remember the distinction between objects and classes.

The OLE lexicon can sometimes confuse the terms *object* and *class*.

## Creating Objects Using MapObjects Object Classes

*Objects must be defined, declared, and instantiated.*

In any object-oriented development environment, creating new objects generally involves a three-step process: *definition*, *declaration*, and *instantiation*. *Define* refers to defining the object class, its properties, and its methods. As you have seen, MapObjects provides (defines) the object classes for you. You use these object classes to create object instances.

*MapObjects supplies the class definitions.*

To create an object instance using a defined class, you must first *declare* the object. In declaring an object, you are stating that the object you are creating will be of the type defined by the object class. After declaring an object, you then *instantiate* it. Instantiation refers to the actual creation of the object through the use of the object class.

*The developer declares and instantiates the object using the class.*

In many languages, including Visual Basic, declaration and instantiation can be performed at the same time. As an example, if you are creating a MapObjects MapLayer object to represent road data, you must declare the object with a name, such as "roadLayer," and instantiate (create) the object as an instance of the MapLayer class. In Visual Basic you would accomplish this with the following line of program code:

```
Dim roadLayer as New MapObjects.MapLayer
```

*In Visual Basic, use Dim to declare and New to instantiate.*

The Dim statement accomplishes the declaration, and the New statement accomplishes the instantiation. Notice that the object type supplied after the New statement is fully qualified. This means that when referencing the MapLayer object class as the object type for this new object, you also specify the object library (in this case, MapObjects) to which the object class belongs. This is not always necessary, but it is good practice because it will prevent any conflicts with duplicate class names provided by other object libraries.

# Accessing Object Methods and Properties

Each of the dozens of GIS and mapping objects delivered by MapObjects has properties and methods that allow a developer and user to interact with it. You may remember from previous chapters that properties are information regarding a particular object, such as the Color property of a Symbol object that indicates what color the symbol should be drawn with. Methods are programmed functions that allow an object to do certain things, such as the ScaleRectangle method of a Rectangle object that tells the rectangle to change its size according to specifications you provide when you invoke the method.

*MapObjects objects incorporate properties and methods.*

To access the methods and properties of objects, you use a format often referred to as *postfix notation.* This simply means that you reference the method or property of an object by supplying the name of the method or property after the object name. The method or property name is separated from the object name by a dot. For instance,

*Properties and methods are accessed with postfix notation.*

```
theObject.theProperty
```

references the property named `theProperty` of the object named `theObject`.

As an example of accessing methods and properties, consider the MapLayer object. Assume you have created a MapLayer named "roadLayer." You can determine whether this layer is visible by retrieving and evaluating its Visible property as follows:

*An example using the Visible property of the MapLayer object.*

```
If roadLayer.Visible Then DoSomeThing
```

Suppose the MapLayer named `roadLayer` has no spatial index and you would like to build one. The MapLayer object provided by MapObjects has a method named Build-Index that you can call to build a spatial index for a MapLayer. To do this you would write the following code:

```
roadLayer.BuildIndex
```

In summary, as a collection of objects made available for use in component-based development environments, MapObjects is representative of GIS software component products. These products follow two important component design priciples: (1) object-orientation, with GIS capabilities provided in the form of objects with properties and methods, and (2) adherence to standard component architectures, such as ActiveX/OLE/COM in the case of MapObjects.

These design principles allow MapObjects and other similar component products to provide GIS capabilities within standard user interfaces such as Windows, and within modern visual programming environments such as Visual Basic. It is these advantages that are enabling a new level of GIS integration in a wide variety of applications through the use of software components.

# 7

# Creating Maps

## MapObjects Maps and Layers

Most GIS software allows graphic display of maps.

A fundamental capability provided by most geographic software is map creation. It is possible to do significant geographic analysis without any graphic display. However, geographic information by its very nature lends itself to visual display. This characteristic of geography, combined with today's graphic-oriented computing environment, has led most developers of geographic software to include some form of graphic map display capability in their products.

This chapter examines objects provided in MapObjects for creating maps.

ESRI's MapObjects, for instance, provides a set of objects that can be used to create and display a wide variety of maps. These objects allow you to create maps that contain multiple pieces of information, such as roads, customer locations, and retail sites; display maps with symbols and colors; and add text and shapes to maps. This chapter examines MapObjects map display objects related to the creation of maps. Map display objects related to symbolization and map rendering are discussed in Chapter 8.

## Maps and Layers

*The Map control accesses objects via events.*

Chapter 6 discussed how MapObjects can be added to a Visual Basic application as an ActiveX control. Regardless of whether your development environment is Visual Basic, adding MapObjects as a component to your application will give you access to the Map control. Once the Map control is available within your application, you can begin adding maps.

## Maps

*A map is where geographic information is displayed.*

In MapObjects, a map is where you display and access your geographic data. It defines a place within your application—on a specific form, for example—in which your map data will display and in which the user can point and click to interact with the map. Note that it is possible to add multiple maps to an application.

### Adding a Map to a Project

Using a Visual Basic project as an example, adding a map to your project is a simple three-step process. A Visual Basic project containing a newly added Map control is shown in the following illustration.

**1.** Open a Visual Basic form.

**2.** Click on the MapObjects control in the toolbox.

**3.** Click and drag a rectangle on the Visual Basic form. Visual Basic places a map on your form with the location and size of the rectangle you clicked and dragged.

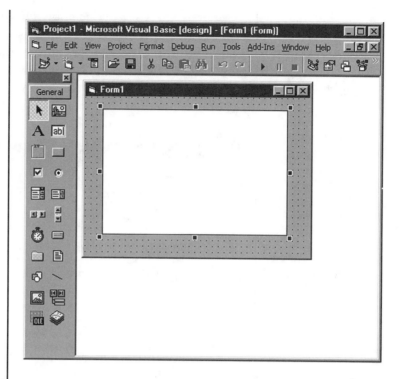

*A Map control added to a Visual Basic project.*

Once you have a map ready for use in your project, the next step is to associate spatial data with the map. To do this requires that you understand Layers and the Layers collection.

## Layers

*Maps have three types of associated layers.*

Maps contain zero or more layers. Layers are collections of geographic data that are thematically related. This means that geographic data contained within a layer are usually similar in spatial representation (e.g., all features in the layer are represented as lines) and content (e.g., all features in the layer represent roads). There are three types of layers available in MapObjects: MapLayers, ImageLayers, and the TrackingLayer.

MapLayers represent vector data.

*MapLayers*, such as the one shown in the following illustration, represent vector data—such as a line map of roads—from a georeferenced data source of a type supported by MapObjects. Currently, the supported formats for MapLayers are ESRI shapefiles, SDE layers, and ARC/INFO coverages.

*A typical MapLayer.*

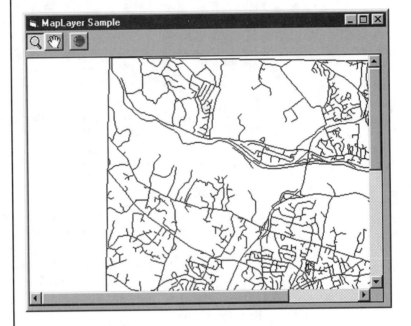

ImageLayers represent raster data.

*ImageLayers*, such as the one shown in the following illustration, represent raster data—such as aerial photography—from a georeferenced data source of a type supported by MapObjects. See the table in Chapter 6 for the currently supported raster formats.

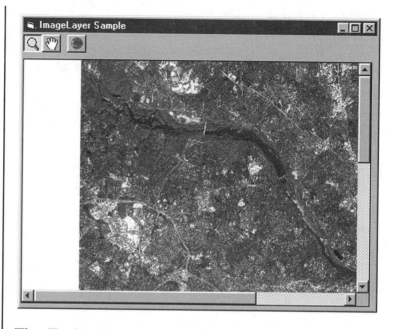

*A typical ImageLayer.*

The *TrackingLayer* is a bit different from MapLayers and ImageLayers. The TrackingLayer provides a layer on which to draw geographic events whose position may change, such as the location of a moving vehicle. To support the dynamic nature of this data, the TrackingLayer has a number of unique methods and properties associated with it. These are discussed later in this chapter, under "The TrackingLayer."

The TrackingLayer represents dynamic data.

## The Layers Collection

A Map's Layers collection stores MapLayers and ImageLayers.

Layers are associated with a Map control through the Layers property of the Map control. The Layers property is defined as a collection, which is an ordered set of related objects. In this case, the set of objects in the Layers collection consists of the layers associated with the map. The layers in a Layers collection may be MapLayers or ImageLayers. The TrackingLayer is not part of the Layers collection, but is identified in a separate property of the Map control.

# Adding Layers to a Map

Layers can be added at design time or run time.

There are two points at which you can add layers to a map: design time or run time. Design time is when you are constructing (designing) your project. Run time is when the project is actually executing (running) as an application. Layers can be added at design time with just a few steps, using the map's Properties window. Adding layers at run time is done by writing program code.

## Adding Layers at Design Time

Add layers at design time if data requirements are fixed.

This method is preferable when you know the specific layers that will be required in the application, and when the location of the files containing the spatial data associated with those layers will remain constant in the computer's file system. Adding layers at design time involves four steps.

1. With your Visual Basic project open, click the right mouse button on the Map control to get a control menu. Select the *Properties...* command to display the Map Control Properties window.

2. Click the Add button on the General tab to display a dialog window that will allow you to navigate your file system and select the data source you want to add to the map's layer collection as a layer. The Files of Type input box on the window will allow you to select from the MapObjects supported file formats for MapLayers and ImageLayers. Locate and select the data source you want to add, then click the Open button.

3. When the dialog window closes you will see the data source you selected added to the Layers list in the Map Control Properties window. Each item in the Layers list represents an individual layer, and the total of all layers listed in the Layers list represents the Layers collection for that map.

**4.** Add as many layers as you like, and close the Map Control Properties window (shown in the following illustration) when you are finished. Clicking on the OK option will save the layer additions you have made; clicking on Cancel option will reject them.

*The Map Control Properties window with a Roads layer added.*

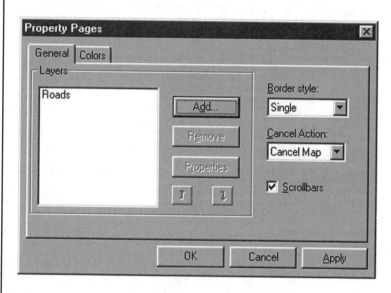

After following these steps with a map, the Layers collection of the map will contain the layers assigned using the Map Control Properties window. Running a project after making these modifications to a map within the project will draw the map layers in the map to which they were assigned.

*New layers are added to a map's Layers collection.*

## Adding Layers at Run Time

*Add layers at run time if data requirements are likely to change.*

Often, you will not know exactly what layers a user will want assigned to a map. You may want to allow the user to remove or add layers to a map when the application is running. You may also encounter situations in which the location of the spatial data files on your computer's file system will not remain consistent. In these situations, you will want to manage layers at run time using program code.

The process for adding layers at run time may vary, depending on the type of data source being accessed. For example, adding a shapefile as a layer requires the use of the DataConnection object, but adding an image does not require this object. The following material examines the process for adding a shapefile as a layer. This is followed by brief examples of adding other data sources in order to highlight differences in the process.

*The process of adding layers varies according to data source type.*

## An Example Using a Shapefile

Programmatically assigning a shapefile as a layer involves the following steps. Each of these steps is discussed in detail in the sections that follow.

*Adding a shapefile as a layer at run time involves programming four steps.*

**1.** Specify a "set" of available data sources from which you will select individual data sources to be assigned to individual layers.

**2.** Create a layer.

**3.** Assign a data source to the layer from the "set" of available data sources.

**4.** Assign the layer to a map.

### Accessing Data Sources with the DataConnection Object

The first step is to establish a set of data sources from which you will later select specific data to assign to specific layers. You must first identify the set of data sources to MapObjects and request MapObjects to establish a connection to this set. You use the DataConnection object to do this. The DataConnection object allows you to specify and connect to an SDE database, ARC/INFO workspace, directory of shapefiles, or other data source supported by MapObjects.

*Step 1: The DataConnection object identifies and connects to a data source.*

*The DataConnection object encapsulates the data source.*

For example, assume you have a directory called *MyShape-Files* located in your file system's root directory and would like to access it using a DataConnection object. To do this, you could write the following code:

```
Dim dConn as New MapObjects.DataConnection
dConn.Database = "C:\MyShapeFiles"
```

The first line creates a new DataConnection object called dConn (recall the discussion of *declaration* and *instantiation* in Chapter 6). The second line sets the Database property of dConn to the path name of your directory of shapefiles. After making a successful connection, the Geo-Datasets property of the DataConnection object will be set to the collection of all data sources available in the source specified by the Database property (in this case, the directory *C:\MyShapeFiles*).

<u>Encapsulation limits the impact of data source changes.</u>

Remember that one benefit of the object-oriented approach of geographic software components is that they can provide data abstraction, as discussed in Chapter 6. In the previous example, the DataConnection object named dConn knows that the physical location of the data is *C:\MyShapeFiles*.

Other objects requiring data do not need to know the physical location; they simply request data from the Data-Connection object, which serves the data to them in response. This means that if the physical location of the data changes, only the DataConnection object must be changed. All other objects are protected from the change by the DataConnection object's encapsulation of the data.

## Creating a Layer

<u>Step 2: A layer is created to receive data from the DataConnection object.</u>

After creating a DataConnection, you have a set of available GeoDatasets in the DataConnection you have established. From this collection of GeoDatasets, you can select a specific data source (a GeoDataset object) for assignment to a layer within a map. However, you must first create that layer. Continuing with the previous example, you might write the following code to create a map layer called MyCity:

```
Dim MyCity as New MapObjects.MapLayer
```

At this point, you have a DataConnection object named dConn, and a MapLayer object named `MyCity`. All that remains is to select a GeoDataset from dConn's GeoDataset collection and assign it to `MyCity`.

### Adding Data to a Layer with the GeoDataset Object

To assign a GeoDataset to a layer, you must search the GeoDatasets collection of your DataConnection in order to find the GeoDataset you want. Assume you have a shapefile named TheCityMap you want to assign to the MapLayer `MyCity`. If this shapefile resides in the directory *C:\MyShapeFiles*, which is the directory to which the dConn DataConnection object is connected, it will exist in the GeoDatasets collection of the dConn DataConnection object. You can locate it in dConn's GeoDatasets collection and make the assignment to the layer using the following code:

*The DataConnection object can contain multiple GeoDatasets.*

```
MyCity.GeoDataset = dConn.FindGeoDataset("TheCityMap")
```

The code on the right side of the equal sign uses the `FindGeoDataset` method of the DataConnection object to locate the shapefile `TheCityMap` within the dConn DataConnection's object's `GeoDataset` collection. The code to the left side of the equal sign takes the GeoDataset returned by the `FindGeoDataset` method (in this case, the GeoDataset named `TheCityMap`) and applies it to the GeoDataset property of the MapLayer object named `MyCity`. The Map layer named `MyCity` will now look to the GeoDataset named `TheCityMap` as its source of data.

*Step 3: The DataConnection assigns one GeoDataset to a layer.*

### Adding a Layer to a Map

Once you have a layer with an assigned GeoDataset, you need to assign that layer to a map. Remember that the map object has an associated Layers collection that is the set of layers assigned to that map. You assign a new layer to a map by adding it to the map's layer collection. If you wanted to assign the layer `MyCity` to a map named Map1, you could use the following line of code:

*Step 4: The layer is added to a map.*

```
Map1.Layers.Add MyCity
```

Add is a method of the Layers collection object that adds the specified layer (in this case, MyCity) to a layer collection.

## Examples Using Other Data Sources

The previous section used the example of a shapefile to review the process of adding a layer at run time. The process varies slightly for other data sources. The following examples highlight some of the variations encountered when adding layers at run time using data sources other than shapefiles. All of the code in these examples is taken from sample Visual Basic projects included on the CD accompanying this book.

### Adding an SDE Layer as a MapLayer

**Adding SDE layers uses access properties of the DataConnection object.**

The following code is taken from the Click event procedure of the Command1 button in Form1 of the sample project named sdeConnect. It demonstrates the use of some properties of the DataConnection objects that were not required in the previous shapefile example.

```
' the values for Server, User, Database, and Password will
' need to be updated for your particular sde connection
dc.Server = "jorel"
dc.User = "sdetest"
dc.Password = "sdbe**"
dc.Database = "Kentucky"
```

In this code, dc is a DataConnection object declared in the General Declarations section of Form1 in the sdeConnect project. The Server property of this DataConnection object specifies the name of the computer on which the SDE database resides. The User and Password properties specify the required user name and password. Recall that when accessing shapefiles, the Database property is set to a directory name. In the case of SDE, the Database property is set to the name of an SDE database, which in this case is Kentucky.

After making the connection, the SDE connection can be used like any other DataConnection object. For example, the FindGeoDataset method can be used to search the DataConnection for specific data (in this case, an SDE layer) to assign to a MapObjects layer.

### Adding a Raster Image as an ImageLayer

The following code is taken from the Form Load procedure in Form1 of the sample project named Image. This code demonstrates how simple it is to add an image as a layer. Note that the DataConnection object is not required.

Adding raster images does not require the DataConnection object.

```
' first add an ImageLayer
Dim iLayer As New ImageLayer
iLayer.File = dataDir & "\Wash.bmp"
Map1.Layers.Add iLayer
```

Notice that the object named `iLayer` is created as an `ImageLayer` rather than a MapLayer, as in the previous examples. Recall that an ImageLayer, although still a layer, is different from a MapLayer. An ImageLayer is for adding raster data to your map. In addition to not having some of the properties found in a MapLayer object, it has an additional property called `File`.

The File property of the ImageLayer identifies the data source.

In this code, the `File` property of the `ImageLayer` object named `iLayer` is set to the full path name of the image. The full path name includes the name of the image, which in this case is `Wash.bmp`. The path name in this code has been previously set in the variable `dataDir`.

After the `File` property of the `ImageLayer` is set to the desired image, the layer can be added to a map's Layers collection just like any other layer. Notice that in working with images, the DataConnection and GeoDataset objects are not required.

## Removing Layers from a Map

Use the Map Control
Properties window
to remove layers
at design time.

Layers are easily removed from a map. Just as layers can be added at design time and run time, they can be removed at design time and run time. To remove layers at design time, as shown in the following illustration, use the same map control properties window used to add a layer at design time. Recall that the Map Control Properties window is accessed by pressing the right mouse button over the Map control and releasing the button over the *Properties...* choice on the menu that appears.

On the General tab, click in the Layers list on the name of the layer to be removed, then click on the Remove button. The layer name will disappear from the Layers list, indicating that the layer has been removed from the layers collection of the map.

*Using the Map Control Properties window to remove a layer at design time.*

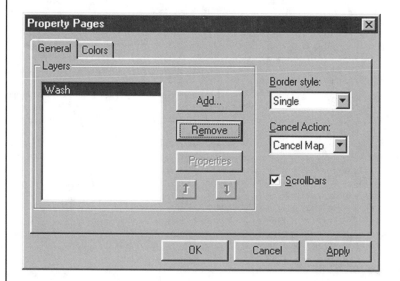

To remove layers at run time, use the Remove method of the Layers collection object. To remove a layer with an index number of 5 (layer indexing is discussed later in

Use the Remove method to remove layers at run time.

this chapter) from the layers collection of a map named Map1, you might write the following code. Removed layers will no longer be accessible via the map.

```
Map1.Layers.Remove 5
```

After following this step with a map, the Layers collection of the map will no longer contain the specified layers.

# Drawing Maps and Layers

This section discusses the conditions under which maps are redrawn and under which layers are visible in drawn maps. It also discusses the order in which layers are made visible (drawn), as well as layer indexing and draw events.

## When Does a Map Redraw?

Map redraws can be implicitly or explicitly caused.

In MapObjects, a map redraw may occur in response to any of several actions. The following actions *implicitly* force a map to be redrawn:

❏ Adding a MapLayer or ImageLayer to a map's Layers collection

❏ Calling the Pan or CenterAt methods of a Map control

❏ Using the scrollbars to pan a map

❏ Updating the Extent property of a Map control

❏ Calling the Clear or Remove methods of the map's Layers collection

You can *explicitly* force a map to be redrawn by using the map's Refresh method as follows:

```
Map1.Refresh
```

## Layer Visibility

All layers have a Visible property that determines whether that layer will be visible when the map is drawn. When a layer is created, MapObjects sets the value of this property to True. This means that the layer will draw when you dis-

The Visible property indicates whether a layer will be drawn.

play the map unless you indicate otherwise by explicitly setting the value of this property to False. For instance, if you had a layer named MyCity in the layers collection of a map named Map1 and did not want that layer to display, you might write the following code:

```
Map1.Layers.Item("MyCity").Visible = False
```

This code makes use of the `Item` method of the layers collection object. The `Item` method requests the collection to return the specified item from its collection. In this case, because the collection is a collection of layers, the item returned is a layer. The line of code shown specifies the name `MyCity` as the name of the layer you want returned. When the layer object named `MyCity` is returned, the `Visible` property is set to the value indicated (`False`).

Code can be simplified by using default methods and properties.

This line of code can be simplified using the concept of default methods. Objects may have default methods and default properties. This means that if parameters are passed to an object without explicit reference to a method or property, the object will accept the parameters and execute the object's default method or set the default property using the parameters. In the case of the Layers collection referenced in the previous line of code, the Item method is the default method of the Layers collection. This means that the code can be rewritten as follows without an explicit reference to the Item method:

```
Map1.Layers("MyCity").Visible = False
```

## Layer Draw Order

Layers are drawn in the order they were added.

When multiple layers are drawn they are drawn in the order in which they were added to the map object. For example, assume the layers MyCountry, MyState, and MyCity are added in that order to a map object. When the map is displayed, MyCountry will be drawn first. MyState will then be drawn on top of MyCountry, and MyCity will be drawn last.

An important consideration when working with ImageLayers is that opaque ImageLayers should be drawn first. Because ImageLayers are often drawn with opaque colors that cover large portions of a map, ImageLayers should generally be added first. This will ensure that the image will draw first and that the other layers will be visible, because they will be drawn *on top of,* rather than underneath, the image.

*Opaque ImageLayers should be drawn first.*

## Layer Indexing

Each individual element in the Layers collection is assigned an index number. The first layer added will be assigned an index number of 0. When a new layer is added, each existing layer's index number is incremented by one, and the newly added layer is then assigned an index number of 0. In the previous example, in which the layers MyCountry, MyState, and MyCity (added to the map in that order) were used, the index numbers would be assigned as follows:

*Index numbers uniquely identify layers in a Layers collection.*

❏   0 = MyCity

❏   1 = MyState

❏   2 = MyCountry

These index numbers can be used to reference individual layers. In the Layer Visibility section you saw the following sample code:

*Index numbers can be used in place of layer names.*

```
Map1.Layers("MyCity").Visible = False
```

Using the appropriate index number, the same code could be rewritten as follows:

```
Map1.Layers(0).Visible = False
```

Because of the way index numbers are incremented, be careful using indexes when layers will be frequently added

Index numbers change when

layers are added.

and removed from a map. For example, if a new layer called MyHome were added to the layer collection for the map in the previous example, the layer collection index numbers would be the following:

❑ 0 = MyHome

❑ 1 = MyCity

❑ 2 = MyState

❑ 3 = MyCountry

Note that the index numbers of all previously existing layers have now been incremented by 1 because of the addition of the new layer to the collection.

## Drawing Events

Drawing a layer triggers a series of Map control events.

When layers are drawn on a map, a series of Map control events occur in the following order. Note that each of these events has an associated procedure in which you can write program code that will execute when the event occurs.

**1.** The BeforeLayerDraw event occurs once for each map layer. It occurs just *prior* to the drawing of that layer. This event is useful for performing actions specific to that layer prior to MapObjects attempting to draw that layer. For example, you may want to place code in this event to check the current map scale and then set the layer's visibility on or off accordingly.

**2.** The AfterLayerDraw event occurs once for each map layer. It occurs just *after* the drawing of that layer.

**3.** The Drawing Cancelled event can occur *anytime* during the map redraw. It is triggered in direct response to the user pressing the Escape key. This is useful for adding any special code you want executed when the user has terminated drawing of the map.

**4.** The BeforeTrackingLayerDraw event occurs once just *prior* to the drawing of the TrackingLayer.

**5.** The AfterTrackingLayerDraw event occurs once just *after* the drawing of the TrackingLayer. Because this is the last drawing event to occur, this event is useful for performing the final activities of your map's display. For example, you may want to draw a shape on top of the map to highlight a selected feature or area.

## Displaying Dynamic Data on the TrackingLayer

*The TrackingLayer displays data the position of which may change.*

*GeoEvents are the data source for the TrackingLayer.*

As previously mentioned, the TrackingLayer has a specific function that is different from that of MapLayers and ImageLayers. Whereas MapLayers and ImageLayers are intended to represent stored data sources that are relatively static (such as disk files), the TrackingLayer is designed to provide a layer on which to draw dynamic geographic data. For example, the location of a moving vehicle can be determined by a Global Positioning System and passed to MapObjects in real time for immediate display on the TrackingLayer.

The GeoDataset property of the MapLayer object and the File property of the ImageLayer object associate these objects with specific data sources. Because TrackingLayer data is usually dynamic (potentially constantly changing), the TrackingLayer does not have properties to associate it with a data source. Instead, the TrackingLayer object has an Event property that stores GeoEvent objects.

## GeoEvent Objects

GeoEvent objects represent dynamic geographic phenomena within the TrackingLayer, such as a vehicle location. GeoEvents have a limited set of properties and methods. An X property stores a horizontal map coordinate, a Y property stores a vertical coordinate, and Sym-

GeoEvents have limited properties.

bolIndex specifies which symbol to use when drawing the GeoEvent on the TrackingLayer. The value in SymbolIndex references a symbol from the TrackingLayer's Symbol property, which stores an array of symbols.

The GeoEvent object has two methods: Move and MoveTo. The Move method moves a GeoEvent relative to its current position, essentially shifting the X and Y coordinates by the amount you specify. The MoveTo method specifies an absolute X and Y map position you want the GeoEvent to move to.

GeoEvents have two movement-related methods.

## The TrackingLayer

TrackingLayers are implicitly created for you.

When you add a map to an application, MapObjects automatically supplies a TrackingLayer for the map. This TrackingLayer can be accessed via the map's TrackingLayer property. This means that you do not need to explicitly create a TrackingLayer; it is already present for you in any map you add.

The AddEvent method creates TrackingLayer GeoEvents.

When you want to add a GeoEvent to the TrackingLayer, use the AddEvent method of the TrackingLayer object. For example, suppose you know that a vehicle's current location in X and Y coordinate space is 1,500 map units (perhaps feet) in the X direction and 2,900 map units in the Y direction. Also suppose that you would like to represent this vehicle location as a point on the TrackingLayer of a map named Map1, using symbol number 3 from the TrackingLayer's Symbol property (an array of symbols). To accomplish this, you might write the following code:

```
Map1.TrackingLayer.AddEvent 1500, 2900, 3
```

The Move and MoveTo methods reposition GeoEvents.

If the vehicle travels an additional 400 feet in the X direction and you want to update the location of the GeoEvent representing that vehicle in the TrackingLayer, you could write the following code:

```
Map1.TrackingLayer.Event(0).Move 400, 0
```

GeoEvents are stored in the TrackingLayer's Event property.

This code references the first, and in this case only, Geo-Event in the TrackingLayer by using the index 0. It then calls the Move method of that GeoEvent, supplying the number 400 for the number of units to move the GeoEvent in the X direction. Any change you make to a GeoEvent will redraw the TrackingLayer of the map; therefore, you do not need to explicitly request a redraw.

It is worth noting that the TrackingLayer has its own Refresh method. This means that the TrackingLayer can be redrawn independent of the other layers of a map. This is especially useful when your TrackingLayer requires frequent update and the other layers of your map contain large amounts of data. In this case, you will not want to redraw the large data layers each time your TrackingLayer updates because the large layers may take some time to draw. The TrackingLayer Refresh method will redraw the TrackingLayer without redrawing the other map layers.

The TrackingLayer can be redrawn independently of other layers.

Refreshes can be limited to changed areas of the TrackingLayer.

A Boolean parameter to the TrackingLayer Refresh method also allows you to limit the refresh of the Tracking-Layer to only those areas that have changed since the last redraw. To refresh only the changed areas of the Tracking-Layer of a map named Map1, you might write the following code:

```
Map1.TrackingLayer.Refresh False
```

In the foregoing code, the word *False* is the Boolean parameter telling the Refresh method not to redraw the entire TrackingLayer, but only the areas updated since the last redraw.

# The Tracking Application: Using Maps and Layers

Among the Visual Basic sample applications on the CD-ROM included with this book you will find a Visual Basic project named Tracking.vbp. If you indicated when installing MapObjects from the CD-ROM that you wanted the samples installed, the Visual Basic applications are available in the *samples\vb* directory located in the directory in which MapObjects was installed on your machine.

This section walks through the Tracking project to demonstrate the use of Maps, Layers, and the TrackingLayer in a working application.

You can simply read this section to gain understanding of the application of Maps, Layers, and the TrackingLayer. However, you are encouraged to open the Tracking project and follow along. In addition to running the code as is, you can gain a better knowledge of the subject matter by changing the source code and then running the code and observing the effects of your changes.

## The Tracking Application's User Interface

The user interface of the Tracking application, shown in the following illustration, is contained in Form1. The interface consists of the following element:

❏   A toolbar (named *Toolbar1* in the Visual Basic project) is supplied to allow the user to specify what actions should occur when clicking on the map with the mouse. The toolbar contains the following buttons:

• A *Zoom* button (Button1 in Toolbar1), displayed with a magnifying glass icon. This button enables the user to zoom in to an area on the map defined by a "clicked and dragged" box.

• A *Pan* button (Button2 in Toolbar1), displayed with a hand icon. This button allows the user to drag the map in any direction to display areas currently outside the bounds of the map.

• An *Add Event* button (Button3 in Toolbar1), displayed with a cross icon. This button adds a new GeoEvent to the TrackingLayer at the location on the map subsequentlyclicked on by the user.

• A *Select Events* button (Button4 in Toolbar1), displayed with a box-and-arrow icon. This button allows the user to select events by clicking and dragging a box to contain events on the map.

The user interface of the Tracking application also contains the following:

❑ A button labeled *Full Extent* (named Command2) will when clicked zoom the map out the furthest extent possible.

❑ A button labeled *Remove Event* (named Command1) will when clicked remove a GeoEvent from the TrackingLayer.

❑ A check box labeled *Data Collection* (named Check1) will when checked cause the events to move randomly on the map.

*The user interface of the Tracking project as seen at run time.*

In addition to the user controls previously described, there are two controls on the form that are visible only in design mode, as shown in the following illustration. An ImageList (named ImageList1) is placed on the form to hold the images to be displayed as icons on the toolbar. A Timer is placed on the form to control the movement of GeoEvents on the TrackingLayer.

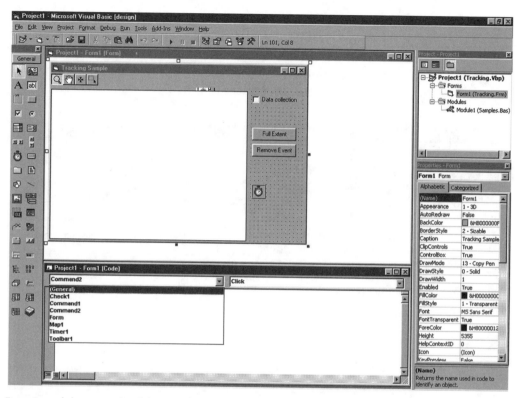

*Form 1 and the controls of the Tracking project displayed at design time.*

## The Form1 Form Load Procedure: Setting Things Up for the User

The Form Load event of Form1 occurs when the Tracking application is started because Form1 is specified as the initial form to load and display when the application begins execution. When the Form 1 Form Load event occurs, the following code, contained in the Form Load procedure, executes:

```
Private Sub Form_Load()
 InitializeMapData
 InitializeTrackingLayer
 Randomize  ' seed the random number generator
End Sub
```

Notice that the code in this procedure calls two procedures: `InitializeMapData` and `InitializeTrackingLayer`. Both procedures are contained in the General section of the Form1 code. The following illustration shows the General procedures of Form1.

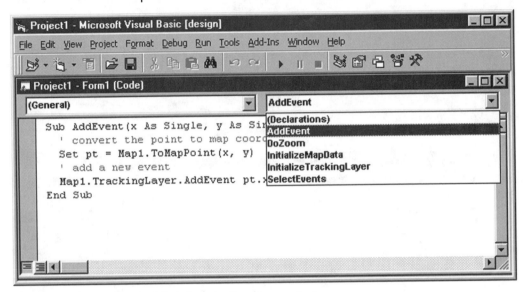

*The General procedures of Form1 in the Tracking project.*

## The InitializeMapData Procedure:
## Adding Layers Using DataConnections and GeoDatasets

The InitalizeMapData procedure called from the Form1 FormLoad procedure contains code to identify data sources and add them to a map as layers. The following is the first block of code in this procedure. It establishes a DataConnection (named dc) and terminates the application if the DataConnection attempt fails, as indicated by the `Connect` property.

```
' load data into the map
Dim dc As New DataConnection
dc.Database = ReturnDataPath("usa")
If Not dc.Connect Then End
```

The following is the second block of code in this procedure. It creates a MapLayer (named "layer") and associates a GeoDataset called "states" (from the DataConnection dc) as the data source for the new layer. The Color property of the layer's Symbol object is then set (use of this object is reviewed in Chapter 8), and the layer is then added to the map named Map1.

```
Dim layer As MapLayer
Set layer = New MapLayer
Set layer.GeoDataset = dc.FindGeoDataset("states")
layer.Symbol.Color = moPaleYellow
Map1.Layers.Add layer
```

The following is the third block of code in the InitializeMapData procedure. It creates another layer to add to Map1. In this case, the GeoDataset assigned to the layer is called ushigh and is assigned a red color.

```
Set layer = New MapLayer
Set layer.GeoDataset = dc.FindGeoDataset("ushigh")
layer.Symbol.Color = moRed
Map1.Layers.Add layer
```

The following is the final block of code in the procedure. It sets the Symbol object of the TrackingLayer to a red circle of size 6. (Chapter 8 contains more on map symbolization and rendering.)

```
' set the symbol of the TrackingLayer
Map1.TrackingLayer.Symbol(0).Style = moCircleMarker
Map1.TrackingLayer.Symbol(0).Color = moRed
Map1.TrackingLayer.Symbol(0).Size = 6
```

## The InitializeTrackingLayer Procedure: Setting Up the TrackingLayer

After the Form1 Form Load procedure executes the InitializeMapData procedure, it calls the InitializeTrackingLayer procedure to create symbols and make them available to the TrackingLayer. The InitializeTrackingLayer contains the following code:

```
Sub InitializeTrackingLayer()
 ' add two symbols to the TrackingLayer for selected
 ' and unselected events

Dim fnt As New StdFont
fnt.Name = "Wingdings"
fnt.Bold = False
Map1.TrackingLayer.SymbolCount = 2

Map1.TrackingLayer.Symbol(0).Color = moBlue
Map1.TrackingLayer.Symbol(0).Style = moTrueTypeMarker
Map1.TrackingLayer.Symbol(0).Font = fnt
Map1.TrackingLayer.Symbol(0).Size = 16
Map1.TrackingLayer.Symbol(0).CharacterIndex = 88

Map1.TrackingLayer.Symbol(1).Color = moDarkGreen
Map1.TrackingLayer.Symbol(1).Style = moTrueTypeMarker
Map1.TrackingLayer.Symbol(1).Font = fnt
Map1.TrackingLayer.Symbol(1).Size = 16
Map1.TrackingLayer.Symbol(1).CharacterIndex = 88
End Sub
```

Because the use of Symbols will be discussed in Chapter 8, it is enough to simply view the code at this point. The effect of the code is to establish two symbols that are available to the TrackingLayer: a blue cross for displaying Geo-Events not currently selected by the user, and a dark green cross for displaying currently selected GeoEvents.

After InitializeMapData and InitializeTrackingLayer execute, the Form1 Form Load procedure executes a Visual Basic Randomize statement to initialize a random number generator for later use. Form1 Form Load procedure then ends, and the application is started and waiting for directions from the user. The following illustration shows how the application looks at start-up.

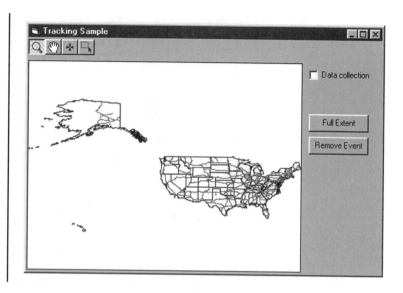

*The user interface of the Tracking project as seen at run time.*

## The Map1 MouseDown Event: Handling a Click on the Map

When the user performs a mouse click on the map, the MouseDown event occurs and the following code from the Map1 MouseDown procedure executes:

```
Private Sub Map1_MouseDown(Button As Integer, Shift As Integer, x As Single, y As Single)
  If Toolbar1.Buttons(1).Value = 1 Then
    DoZoom Shift
  ElseIf Toolbar1.Buttons(2).Value = 1 Then
    Map1.Pan
  ElseIf Toolbar1.Buttons(3).Value = 1 Then
    AddEvent x, y
  ElseIf Toolbar1.Buttons(4).Value = 1 Then
    SelectEvents
  End If
End Sub
```

Notice first that the MouseDown event is defined by MapObjects to receive four parameters: an integer value named `Button`, an integer value `Shift`, and two single precision numbers named `x` and `y`. See Chapter 6 for a list of all events recognized by MapObjects and explanations of their available arguments.

The If statements in the Map1 MouseDown procedure determine what button is currently selected in the toolbar. For example, if button 1 (the "Zoom In" button) is selected, the DoZoom procedure is executed, passing the value contained in the Shift variable that is defined as an argument of the MouseDown event.

### The Toolbar 1 Zoom Button: Using the Map Extent to Zoom In and Out

Recall from Chapter 6 that the value contained in the Shift variable is set by MapObjects when the mouse is clicked on the map. This value indicates whether a combination of the Shift, Alt, and Ctrl keys was pressed when the map was clicked. If no combination of these keys was pressed when the MouseDown event occurred, the Shift variable will have a value of 0. Otherwise, the Shift variable will contain a non-zero value. The DoZoom procedure contains the following program code:

```
Sub DoZoom(Shift As Integer)
  If Shift = 0 Then
   Map1.Extent = Map1.TrackRectangle
  Else
   Set r = Map1.Extent
   r.ScaleRectangle 1.5
   Map1.Extent = r
  End If
End Sub
```

This code receives the Shift argument passed to it. If the value is 0 (no combination of the Shift, Alt, and Ctrl keys was pressed), the Map's Extent property is set to a new rectangle defined by the user. The Extent property of a map defines the dimensions of the area that will be displayed on the map. The Extent property must be set to a

Rectangle object, which is a MapObjects geometric object representing a box.

In this code, the `TrackRectangle` method of the map is called to allow the user to define a rectangle by clicking and dragging on the map. When the user completes the click-and-drag operation, the `TrackRectangle` method returns a Rectangle object that is used to update the map's extent. This forces a map refresh, and the result is a map zoomed in to the area defined by the user-defined box, as shown in the following illustration.

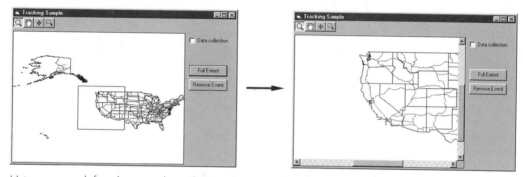

*Using a user-defined rectangle in the Tracking application for zooming in on a map.*

If the value of the `Shift` variable is non-zero (some combination of the Shift, Alt, and Ctrl keys is pressed), the code after the `Else` statement is executed to accomplish a zoom *out*. This is done by setting a variable (named `r`) to contain the Rectangle object returned from the map's Extent property. The `ScaleRectangle` method of the Rectangle object is then called to increase the size of the Rectangle object by a factor of 1.5. The map's Extent property is then updated to contain the newly resized Rectangle, `r`. The result is a map zoomed out to an area 1.5 times larger than the previous map display.

## *The Toolbar 1 Pan Button: Using the Map.Pan Method*

Now return to the If statement in the MouseDown event procedure of Map1 to examine what happens if button 2 in the toolbar (the Pan button) was the selected button when the mouse-click event on the map triggered Map1's MouseDown event. If button 2 was selected, the Pan method of the map object is called using the following code to convert from control units to map units:

```
Map1.Pan
```

Pan is a method of the MapObjects Map object that allows the user to move the map by clicking on the map and then moving the mouse while holding the mouse key down. This will pan the map (move it in any direction) according to the user's mouse movements.

## *The Toolbar 1 Add Events Button: Adding GeoEvents to the TrackingLayer*

If button 3 in the toolbar (the "Add Events" button) is the selected button when the MouseDown event occurs, the AddEvents procedure is executed. Two values, contained in the variables $x$ and $y$, are passed to the AddEvents procedure when it is called. This is done in the following code from the MouseDown event procedure:

```
ElseIf Toolbar1.Buttons(3).Value = 1 Then
  AddEvent x, y
```

Remember that x and y are parameters the MapObjects Map control automatically passes to the MouseDown event. The x parameter contains the value of the X coordinate of the location of the mouse click on the Map control. The y parameter contains the value of the Y coordinate.

Note that these coordinates are expressed in control units: the unit of measurement used in your Visual Basic environment, such as twips. Twips are a screen-independent unit used by Visual Basic to ensure that graphic displays generated by an application are placed and proportioned the same on all display systems. There are 1,440 twips per inch. The AddEvents procedure contains the following code:

```
Sub AddEvent(x As Single, y As Single)
 ' convert the point to map coordinates
 Set pt = Map1.ToMapPoint(x, y)
 ' add a new event
 Map1.TrackingLayer.AddEvent pt.x, pt.y, 0
End Sub
```

This simple procedure first receives the values x and y from the MouseDown event, then uses the ToMapPoint method of the MapObjects Map object to convert the x and y values from control units into map units (e.g., feet). ToMapPoint returns a MapObjects Point, a geometric shape with a single pair of X and Y coordinate values. Therefore, the net effect of the line containing the ToMapPoint reference is to convert the coordinates from control units to the real-world units of your map, then convert the coordinate pair to a Point object referenced by the variable pt.

The AddEvent method of the TrackingLayer is then called to add a GeoEvent to the TrackingLayer of Map1. Notice that the values of the X and Y coordinates of the new point object are retrieved by referencing the point object's X and Y properties. These values are passed—along with a reference to symbol 0 as the symbol with which to draw the Geo-Event—to the AddEvent method of the TrackingLayer. The addition of the GeoEvent forces a refresh of the TrackingLayer, and the GeoEvent appears on Map1. The result of this process is shown in the following illustration.

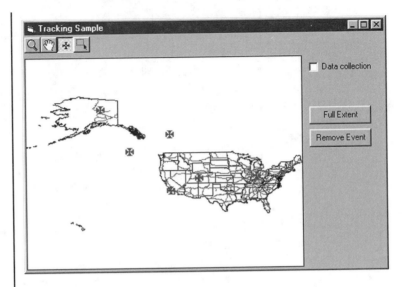

*The Tracking application after the addition of several GeoEvents (designated by the cross symbols).*

## The Toolbar 1 Select Events Button: Accessing GeoEvents on the TrackingLayer

Returning to the Form1 Map1 MouseDown event procedure, if button 4 (the "Select Events" button) were the selected button when the MouseDown event occurred, the SelectEvents procedure would be called. The SelectEvents procedure contains the following code:

```
Sub SelectEvents()
 Set r = Map1.TrackRectangle
 nEventCount = Map1.TrackingLayer.EventCount
 Dim testPt As New Point
 For i = 0 To nEventCount - 1
  Set evt = Map1.TrackingLayer.Event(i)
  testPt.x = evt.x
  testPt.y = evt.y
  If r.IsPointIn(testPt) Then
   evt.symbolIndex = 1
  Else
   evt.symbolIndex = 0
  End If
```

```
  Next i
End Sub
```

The first three lines in the procedure accomplish the following: (1) a variable, r, is set to contain a rectangle defined by the user using the TrackRectangle method of the Map object, (2) using the TrackingLayer method EventCount, a variable, nEventCount, is set to the number of GeoEvents currently present in the TrackingLayer of Map1, and (3) a variable, testPt, is declared to be used as a MapObject's Point object.

The For loop (the code contained between For and Next) executes once for each GeoEvent. This is controlled by the For statement:

```
For i = 0 To nEventCount - 1
```

This statement indicates that a counter named i will begin with a value of zero and will be incremented by 1 each time the block executes, and that execution will stop when the value exceeds the number of events minus 1 (nEventCount - 1).

A range beginning with 0 is used to allow you to use i as an index number to reference each GeoEvent sequentially, one in each execution of the block. Beginning at 0 is necessary because index numbers for GeoEvents in the GeoEvents collection begin with 0, not 1.

The variable evt is set to the current GeoEvent object, and the X and Y properties of the testPt object are set to the X and Y values of that GeoEvent. The If statement (the code between If and End If) then executes to determine if testPt (which is now equivalent in location to the current GeoEvent) falls within the user-defined rectangle contained in the variable r. The IsPointIn method is used for this.

IsPointIn is a method made available by polygonal geometric objects, such as rectangles and ellipses, defined by

MapObjects. IsPointIn returns a Boolean indicating whether a specified Point object is located within the specified polygonal object. The creation of a Point object equivalent to the GeoEvent object was necessary because IsPointIn requires a point (rather than a GeoEvent) as its argument.

If the point (equivalent to the GeoEvent) falls within the rectangle, IsPointIn returns True, the If statement evaluates to True, and the symbol of the current GeoEvent is set to 1 using the SymbolIndex property of the GeoEvent. If the point does not fall within the rectangle, IsPointIn returns False, the If statement evaluates to False, and the code after the Else statement executes. This code sets the symbol of the current GeoEvent to 0 using the GeoEvent's SymbolIndex property.

The entire procedure ends execution when all GeoEvents in the TrackingLayer have been processed. After the TrackingLayer is refreshed the final time, the result is a map displaying the user-selected GeoEvents with symbol 1, and all other GeoEvents with symbol 0.

## The Command1 Button: Removing GeoEvents from the TrackingLayer

The Command1 button (labeled "Remove Event" in the user interface) will when pressed remove an event from the TrackingLayer. The program code to accomplish this is contained in the Command1 button's Click procedure. When the Command1 button is pressed, the button's Click event occurs. When the Click event occurs, the Command1 Click procedure runs and executes the following program code:

```
Private Sub Command1_Click()
 ' remove the first event
 If Map1.TrackingLayer.EventCount > 0 Then
  Map1.TrackingLayer.RemoveEvent 0
 End If
End Sub
```

This procedure first uses the EventCount property of the TrackingLayer to determine if there are any GeoEvents located on the TrackingLayer of Map1. If the value returned by EventCount is greater than 0, GeoEvents exist and the RemoveEvent method of the Tracking-Layer is called using a GeoEvent index of 0. This removes the GeoEvent at index 0, which is the first event in the list. If there are no GeoEvents in the TrackingLayer of Map1, this procedure will have no effect.

## The Command2 Button: A View of All Map Layers

The Command2 button (labeled "Full Extent" in the user interface) will when pressed zoom the map display out to the farthest extent possible. This will provide a view of all Map1 map layers and features. The program code for this is contained in the Command2 button's Click procedure. When the Command2 button is pressed, the button's Click event occurs. When the Click event occurs, the Command2 Click procedure runs and executes the following program code:

```
Private Sub Command2_Click()
 ' zoom to full extent
 Map1.Extent = Map1.FullExtent
End Sub
```

The single line of code in this procedure sets the map's Extent property (a rectangle representing the currently visible portion of the map) to the value of Map1's FullExtent property (the minimum-size rectangle that will include all features on all layers within the map). Changing the Extent property forces a refresh, and the result is a map zoomed out to its maximum extent.

## The Check1 Check Box: Moving GeoEvents on the TrackingLayer

The Check1 check box (labeled "Data Collection" in the user interface) will when checked by a user mouse click begin to randomly move the GeoEvents on the Tracking-Layer. This illustrates the dynamic nature of the Tracking-Layer and simulates real-time location acquisition through devices such as Global Positioning System receivers. When unchecked by a user mouse click, the movement of the GeoEvents is stopped.

Program code associated with the Check1 check box is contained in the check box's Click procedure. When Check1 is clicked, the check box's Click event occurs. When the Click event occurs, the Check1 Click procedure runs and executes the following program code:

```
Private Sub Check1_Click()
 ' toggle on/off the timer
 If Check1.Value = 0 Then
  Timer1.Interval = 0
 Else
  Timer1.Interval = 500
 End If
End Sub
```

This procedure serves a single purpose: to turn a Timer object on or off depending on the value of the check box. If the user click removes the check mark from the check box (indicated by the check box's Value property being set to 0), the Timer is turned off; otherwise, it is turned on.

A Visual Basic Timer control recognizes a single event: elapsed time (as indicated in the timer's Interval property). The Interval property of a Timer control is specified in milliseconds. This means that an interval of 0 effectively disables the Timer control, and an interval of 500 causes the Timer's Timer event to occur every 500 milliseconds, or .5 seconds.

When the check box (the Check1 check box) on the form is checked by the user (indicated by the check box's Value property being set to 1), indicating that data collection should begin, the Timer's Interval property is set to 500 milliseconds and the following code, contained in the Timer's Timer event procedure, begins executing once every 500 milliseconds, or .5 seconds.

```
Private Sub Timer1_Timer()
 maxDist = Map1.Extent.Width / 20
 nEventCount = Map1.TrackingLayer.EventCount
 For iIndex = 0 To nEventCount - 1
   Set gEvt = Map1.TrackingLayer.Event(iIndex)
   gEvt.Move maxDist * (Rnd - 0.5), maxDist * (Rnd - 0.5)
 Next iIndex
End Sub
```

The first line sets a variable, maxDist, to a distance, in map units, equal to 1/20 the horizontal dimension of the current Extent of Map1. The second line sets a variable, nEventCount, to the number of GeoEvents in the TrackingLayer of Map1.

The For loop is executed once for each GeoEvent in the TrackingLayer in a manner similar to that of the previously discussed SelectEvents procedure. The first line of the For loop sets a variable, gEvt, to the current GeoEvent. The next line, which follows, is where the actual move of the GeoEvent occurs.

```
gEvt.Move maxDist * (Rnd - 0.5), maxDist * (Rnd - 0.5)
```

The Move method of the GeoEvent object is called with two values representing the X and Y coordinates of the new location to which the GeoEvent should be moved. The X is a value of maxDist multiplied by the value of a random number (generated using the Visual Basic Rnd function) minus .5. The Y value is calculated in the same manner. Recall that invoking the Move method forces a redraw of the TrackingLayer. The redrawing of the TrackingLayer after each movement of a GeoEvent produces the animation effect desired in real-time location tracking.

The procedure ends execution after all GeoEvents have been processed. Because this procedure is contained in the Timer event of a Timer with an Interval property set to 500, the procedure executes once every .5 seconds until the Timer is turned off by a click on the Check1 check box.

# 8

# Displaying Maps

## MapObjects Symbols and Renderers

GIS components allow maps to be displayed in non-GIS applications.

The previous chapter discussed and demonstrated how MapObjects is used to create maps. However, creating a map is only the first step in the process of communicating geographic information in visual form. How a map is displayed determines the real effectiveness with which the map visually communicates information to its user. Various implementations of geographic software components provide varying degrees of support for manipulation of the map display. This support can range anywhere from the creation of "canned" standard maps to fully customizable maps.

MapObjects provides Symbol and Renderer objects for map display.

It is often necessary to use such elements as classifications, symbols, colors, and text to highlight and contrast various aspects of mapped information. A geographic software component can supply these elements as objects for use by the component user. MapObjects does just this by providing two sets of objects: Symbols and Renderers.

# Drawing Layers with Symbols

Symbols are the *graphic characteristics* of the *geometric* features drawn on a map to represent the *geographic* features of a layer. For example, a road (a *geographic* feature) may be represented on a map as a line (a *geometric* feature) that is dashed and has a specific width (*graphic characteristics*).

Symbols represent graphic characteristics of features.

The graphic characteristics of a geometric feature must exist independent of that feature in order to provide maximum flexibility. This allows the graphic characteristics to be modified without modifying the geometric feature itself. In the case of the previous example, you might want to change the dashed, wide line representing a road to a dotted, narrow line. With graphic characteristics existing as an object distinct from the feature, this can be done by modifying the graphic characteristics object associated with the road feature rather than the road feature itself.

Symbol objects are distinct from features.

In MapObjects, the object representing graphic characteristics is the Symbol object. The Symbol object is independent of a feature but can be associated with a feature. In addition to this independence allowing the Symbol to be modified independent of the features, it also allows a Symbol object to be defined once and then associated with multiple geometric objects.

A single Symbol can be associated with multiple features.

## Modifying Symbols at Design Time

At design time, use the Map Control Properties window to select symbols.

Symbols may be assigned to layers when you are designing your application. In the previous chapter, you learned how layers can be added at design time by using the Map Control Properties window. Remember that this window is accessed by clicking the right mouse key on the Map control window.

If layers have been added to the map at design time, they will display in the layers list box in the Map Control Properties window. To assign a symbol to any of these layers, select the layer and then press the Properties button in the Map Control Properties window. This will display the Layer Properties window for that layer, as shown in the following illustration.

*Accessing the Symbol properties of Color, Size, and Style at design time for a point layer.*

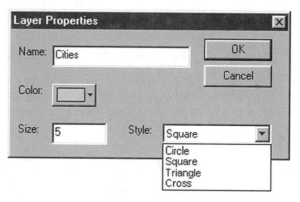

Select from symbol styles relevant to the feature type (i.e., point, line, or polygon).

The Layer Properties window allows you to change the color, size (in typesetting points, with a point being 1/72 of an inch), and style of the symbol associated with the layer. The options will change according to the types of features in the layer (i.e., points, lines, or polygons).

For instance, if the layer for which symbol properties are being modified is a polygon layer, the style options will include "Upward diagonal" for a hatching pattern to fill the polygon. However, if the layer consists of points, the style options will not include "Upward diagonal" because that is not applicable to a point. The new style options presented will include Square, Triangle, Circle, and Cross to indicate the shapes possible for displaying a geometric point.

To accept modifications made to the layer's symbol, click OK in the Layer Properties window, then click OK in the Map Control Properties window. The symbol definition you specified using the Layer Properties window will now be used when MapObjects draws the associated layer.

## Modifying Symbols at Run Time

When layers are added to an application—whether at design time or run time—MapObjects will automatically associate a Symbol object with the layer and assign default values to the Symbol object's properties. Sometimes you may want the symbol definitions or the symbol-to-layer associations changed while the program is running.

For instance, you may want to give the user the ability to modify the symbolization of a layer. To do this involves manipulating symbols and their associations at run time. The following basic example of run-time symbol modifications involves the same properties that can be modified at design time using the Layer Properties window: Color, Size, and Style.

Suppose you have a layer named MyCities containing point features representing cities. If you want all points in the layer to be drawn as yellow circles 5 points in size, you could write the following code:

```
Map1.Layers("MyCities").Symbol.Color = moYellow
Map1.Layers("MyCities").Symbol.Style = moCircleMarker
Map1.Layers("MyCities").Symbol.Size = 5
```

Notice the use of `moYellow` to designate the color and `moCircleMarker` to specify that a circle be used as the symbol. These are MapObjects constants. MapObjects supplies a wide variety of constants to provide predefined values for various purposes. For example, MousePointer constants such as moArrow and moPencil are supplied to provide predefined icons for use in displaying the mouse pointer. For working with symbols, MapObjects provides color constants and style constants for indicating predefined colors and symbol styles that can be applied to a Symbol object.

## Colors and Color Constants in MapObjects

*There are many ways to identify a specific color.*

Colors in MapObjects may be specified in many ways: as a long integer providing values for red, green, and blue (RGB); a color constant provided by the programming environment, such as vbGreen in Visual Basic; by using functions provided by the programming environment to return color values, such as the RGB or QBColor functions in Visual Basic; or with the MapObjects color constants, such as moYellow. Very often, the same color can be specified in several of these ways.

*An easy way to specify colors is to use the MapObjects color constants.*

MapObjects color constants are a set of about two dozen standard colors identified by a constant name. For example, moBlack, moRed, and moGreen are MapObjects color constants. Notice the prefix "mo" to distinguish them as MapObjects (mo) constants. For a complete list of the MapObjects color constants, see the MapObjects help files on the CD-ROM accompanying this book.

## Symbol Style Constants in MapObjects

*The types of features in a layer determine the symbol type.*

The style constants available for use with a given Symbol object are dependent on the SymbolType of the Symbol object. SymbolType is a property of a Symbol object, with possible values of moPointSymbol, moLineSymbol, and moFillSymbol.

*Points, lines, and polygons each have a symbol type.*

When a MapLayer is created, MapObjects automatically sets the SymbolType property of a layer's symbol according to the type of features in the layer. If a layer consists of point features, its Symbol object's SymbolType property is assigned a value of moPointSymbol. If a layer consists of line features, SymbolType is set to moLineSymbol. A layer with polygons will have SymbolType set to moFillSymbol (for "filling" the polygon with a pattern or color).

Each symbol type has many possible symbol styles.

A given symbol type may have many different symbol styles. For instance, a point symbol (SymbolType of moPointSymbol) may be styled as a circle, a square, or a triangle. The value of the SymbolStyle property of the Symbol object indicates the style. SymbolStyle constants are provided as a means of accessing predefined symbology, such as moCircleMarker (a circle) or moSquareMarker (a square). Each SymbolType has a set of SymbolStyle constants associated with it, as follows. The illustration that follows shows an example of the use of symbol types and styles.

❐ SymbolType *moPointSymbol* has the following SymbolStyles:

> *moCircleMarker* for a circle
> *moSquareMarker* for a square
> *moTriangleMarker* for a triangle
> *moCrossMarker* for a cross
> *moTrueTypeMarker* for a character from a TrueType font

❐ SymbolType *moLineSymbol* has the following SymbolStyles:

> *moSolidLine* for a solid line
> *moDashLine* for a dashed line
> *moDotLine* for a dotted line
> *moDashDotLine* for a dash-dot line
> *moDashDotDotlLine* for a dash-dot-dot line

❐ SymbolType *moFillSymbol* has the following SymbolStyles:

> *moSolidFill* for solid
> *moTransparentFill* for transparent
> *moHorizontalFill* for horizontal lines

*moVerticalFill* for vertical lines
*moUpwardDiagonalFill* for upward diagonal lines
*moDownwardDiagonalFill* for downward diagonal lines
*moCrossFill* for cross hatching
*moDiagonalCrossFill* for diagonal cross hatching
*moLightGrayFill* for light gray fill
*moGrayFill* for gray fill
*moDarkGrayFill* for dark gray fill

In the following illustration, the states of Mexico are drawn with SymbolType moFillSymbol and SymbolStyle moSolid-Fill. Roads are drawn with SymbolType moLineSymbol and SymbolStyle moDashLine. Cities are drawn with Symbol-Type moPointSymbol and SymbolStyle moCrossMarker.

*An example of the use of symbol types and styles.*

Note that the moTrueTypeMarker SymbolStyle can be used to indicate the use of a specific character from a TrueType font as the graphic for a symbol. For example, you may want to use the letter H as a marker icon for points representing hospitals. TrueType fonts are scaleable fonts, which means

Characters from TrueType
fonts can be used
as symbols.

that they produce a high-quality character regardless of font size.

When using moTrueTypeMarker, two other properties of the Symbol object become relevant: CharacterIndex and Font. Font specifies the TrueType font from which to choose a character, and CharacterIndex specifies the index number of the character to use from the font's character set. The Symbol object also supports properties for specifying the angle at which a symbol should be drawn (the Rotation property), whether outlines should be drawn for polygons (the Outline property), and the color of polygon outlines (the OutlineColor property).

## The TextSymbol Object

*TextSymbol objects draw text for a feature.*

The TextSymbol object is an object used to symbolize text, and exists as an object distinct from the Symbol object because of the unique requirements for symbolizing text. For instance, the TextSymbol object, as shown in the following illustration, has a Fitted property to contain a value indicating whether to adjust the gap between characters of a TextSymbol object so that it fits between two points of a line. This is not required when symbolizing a feature, but can be very useful when attempting to position text.

*An example of the use of the TextSymbol object. This fit of the text to the line is affected by the value of the Fitted property of the TextSymbol object used to draw the text.*

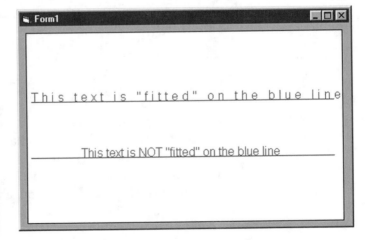

# *Drawing Features with Renderers*

You may have realized by now that by manipulating Symbol and Layer object associations you are really only manipulating symbols at the layer level. This means that all features in a layer will be drawn with the same symbol, specified by the Symbol object associated with the layer.

There may be occasions when you need to symbolize individual features within a layer differently. This can occur when you want to highlight features according to how their attribute values are distributed among value ranges. For example, you may want to use blue-colored points to represent all cities with a population in the range 0 to 500,000, and red-colored points to represent cities with a population greater than 500,000. MapObjects provides Renderer objects to accommodate requirements such as this.

The following descriptions of Renderer objects summarize the purpose of each type of renderer provided by MapObjects. More detailed discussion of the renderers and their properties and methods is provided in the application walkthrough comprising the last section of this chapter.

## The DotDensityRenderer: Symbolizing with Density Patterns

If you have a polygon layer with an attribute to which density is relevant, a dot density map can provide a visual display of regions of increasing or decreasing density. Population maps often use this technique because density is an important aspect of population distribution.

In a dot density map used to represent population, such as the one shown in the following illustration, polygons with higher populations will have more dots than polygons with lower populations. The resulting effect is one of a higher density of dots in regions with a higher density of population. To allow the creation of dot density maps, MapObjects provides the DotDensityRenderer object.

*A population density map of the northeastern United States created with the DotDensityRenderer.*

## The ValueMapRenderer: Symbolizing by Attribute Values

*The ValueMapRenderer symbolizes unique attribute values.*

There are occasions when you may want to symbolize the features on your map based on their unique attribute values. For instance, you may have a map of polygons that represent city zoning designations. One way to present this map would be to label each polygon with the zoning code so that a user of the map could look at a polygon and see its zoning designation by reading its label.

However, a more useful way to present the map would be to shade each polygon with a color based on its zoning designation. With this approach, every polygon with a zoning of "residential" might be colored blue, and every polygon with a zoning of "commercial" might be colored green.

In addition to being able to interpret the zoning of any polygon by simply noting its color, a user of this map could see the distribution of zoning types by noting the distribution of color. To produce maps that are symbolized by unique attribute values, such as the one shown in the following illustration, MapObjects provides the ValueMapRenderer object.

*A map created with the ValueMapRenderer symbolizing the "state" attribute for each county in the northeastern United States.*

## The ClassBreaksRenderer: Symbolizing by Classification

*The ClassBreaksRenderer symbolizes ranges of attribute values.*

In some situations it is useful to symbolize the features on a map based on attribute value ranges. In contrast to a ValueMapRenderer map based on unique attribute values, this type of map uses the same symbol to depict all features with values for a given attribute that fall within a specified range for that attribute. This means that although many unique values may be contained in a single range, only one Symbol will be used to represent all features in that range, regardless of their individual values.

This method is commonly used in displaying maps of population. If you have a map for which population counts exist (such as census tracts), you may want to display polygons with a high population in red, polygons with a moderate population in blue, and polygons with a low population in green. To make such a map, you would define ranges of population values for the high, medium, and low classifications and then apply a symbol to each range. To create maps that are symbolized by ranges of attribute values, such as the one shown in the following illustration, MapObjects provides the ClassBreaksRenderer object.

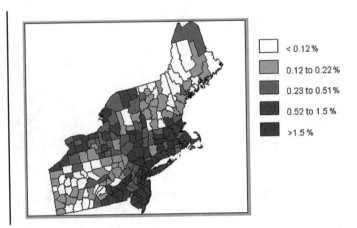

*A map created with the ClassBreaksRenderer displaying the percentage of each county's population of unreported ethnicity per the 1990 U.S. Census.*

## The LabelRenderer: Labeling a Map with Text

*The LabelRenderer symbolizes using text created from attribute values.*

In some cases you may not want to symbolize your map with color values or symbols that are markers, shade patterns, or the like. Instead, you may want to label features with text that displays an attribute value.

For instance, you may have a map of school locations and you may want to place a label next to each point representing a school. This label could display the value of the field containing the name of the school. To make maps with features that are labeled with attribute values, such the one shown in the following illustration, MapObjects provides the LabelRenderer object.

*A map created with the LabelRenderer, with each county in northern Maine labeled with its name.*

## The ThematicMap Application: Using Symbols and Renderers

Among the Visual Basic sample applications on the CD-ROM included with this book, you will find a Visual Basic project named ThematicMap.vbp. If you indicated when installing MapObjects from the CD-ROM that you wanted the samples installed, the Visual Basic applications are available in the *samples\vb* directory located in the directory in which MapObjects was installed on your machine. The ThematicMap project demonstrates the use of renderer objects to display a map of the northeastern United States in several ways. Each renderer is used with a variety of symbols and data to create a series of thematic maps for the region. This section walks through the ThematicMap project to examine Symbol and Renderer objects as they are used in an actual application.

Although you can simply read this section to gain an understanding of the application of Symbols and Renderers, you are encouraged to open the project and follow along. You can run the code as it stands, and because the source code is provided, you can get hands-on experience by changing values and running the code to see the effect of your changes.

## The ThematicMap Application's User Interface

The user interface of the ThematicMap application is contained in Form1. The interface consists of the following elements. The illustrations that follow show the ThematicMap project at run time and at design time.

❑ A *MapObjects Map* control (named Map1 in the Visual Basic project) that provides the display area for drawing the map

❑ A *Single Symbol Map* button (named Command2) that when pressed will draw all features on the map using the same symbol

❑ A *Dot Density Map* button (named Command1) that when pressed will create a dot density map

☐ A *Value Map* button (named Command3) that when pressed will display the features on the map with symbols based on unique attribute values

☐ A *Standard Deviation Map* button (named Command4) that when pressed will display the features on the map with symbols based on a calculation of standard deviation

☐ A *Quantile Classes Map* button (named Command8) that when pressed will display the features on the map with symbols based on a classification that ensures the same number of features will be contained in each class range

☐ A *Graduated Symbol Map* button (named Command5) that when pressed will display the features on the map with symbols that increase in size as the value of a specified attribute increases

☐ A *Label Map* button (named Command6) that when pressed will label each feature on the map with a value from a specified attribute

☐ A *Full Extent* button (named Command7) that when pressed zooms the map out to the maximum extent

*The user interface of the ThematicMap project as seen at run time.*

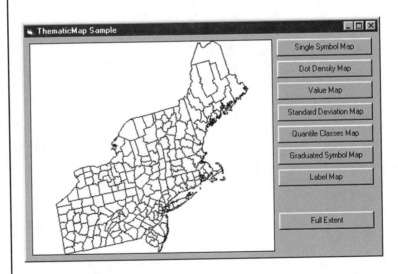

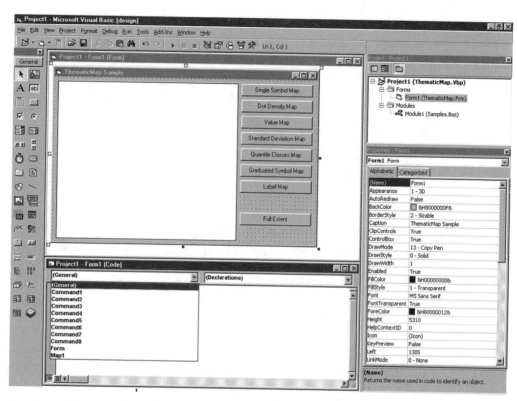

*Form1 and the controls of the ThematicMap project displayed at design time.*

## The Form1 Form Load Procedure: Setting Up Maps and Symbols

The Load event of Form1 occurs when the ThematicMap application is started. This is because the ThematicMap project is set to automatically load and display Form1 when executed. This causes the form's Load event to occur. When this event occurs in the ThematicMap application, the following program code, contained in Form1's Form Load procedure, is executed:

```
Private Sub Form_Load()
  ' load data into the map
  Dim dc As New DataConnection
  dc.Database = ReturnDataPath("Northeast")
  If Not dc.Connect Then End
```

```
Dim layer As MapLayer
Set layer = New MapLayer
Set layer.GeoDataset = dc.FindGeoDataset("Counties")
layer.Symbol.Color = moPaleYellow
Map1.Layers.Add layer


Set layer = New MapLayer
Set layer.GeoDataset = dc.FindGeoDataset("NeCenter")
layer.Visible = False
Map1.Layers.Add layer
End Sub
```

The first block of code in this procedure creates a `Data-Connection` object and points it to a directory containing shapefiles. If a valid connection to the directory is not made, the application will exit, as indicated by the line `If Not dc.Connect Then End.`

The second block of code in the procedure creates a `MapLayer` and assigns a shapefile named `Counties` as the `MapLayer`'s GeoDataset. (The use of DataConnection objects and MapLayer objects was discussed and demonstrated in Chapter 7.) Note the second to last line of the second block of code:

```
layer.Symbol.Color = moPaleYellow
```

This line changes the color of the MapLayer's symbol. To do this, the code references the `Color` property of the `Symbol` object associated with the object named `layer`, which is the MapLayer referencing the Counties shapefile. The `= moPaleYellow` portion of the line sets the `Color` property to the value of the MapObjects constant `moPaleYellow`. This constant represents the color pale yellow. The final line in the second block of code adds the MapLayer to the map named `Map1`.

The third and final block of code in the procedure adds one more MapLayer to the map. The shapefile data source for this layer is `NeCenter`, a map of points representing the center of each county. `NeCenter` will be used for only one type of map, the Graduated Symbol map. Therefore, its `Visible` property is set to `False` (off) to make it invisible (not drawn). When it is required for the Graduated Symbol map, the visibility property will be reset to True.

## The Map1 MouseDown Event: Handling a Click on the Map

When the user clicks the mouse on the map named Map1, Map1's MouseDown event occurs. When the MouseDown event occurs, the following code from the Map1 Mouse-Down procedure is executed:

```
Private Sub Map1_MouseDown(Button As Integer, Shift As Integer, x As Single, y As
Single)
   If Button = 1 Then
      ' zoom in
      Set r = Map1.TrackRectangle
      If Not r Is Nothing Then Map1.Extent = r
   Else
      ' zoom out
      Set r = Map1.Extent
      r.ScaleRectangle 1.5
      Map1.Extent = r
   End If
End Sub
```

In the first line of the procedure, the `If Button = 1` statement checks to see whether or not the mouse button used to perform the mouse click was the number `1` (left) mouse button. If it was, a zoom-in is performed. To perform the zoom-in, the variable `r` is set to reference a Rectangle object returned from the `TrackRectangle` method.

Recall from the Tracking application example in Chapter 7 that the TrackRectangle method allows the user to click and drag a rectangle shape on the map. This rectangle will designate the new map extent (the area the user wants displayed on the map).

The final line in this portion of the If block is:

```
If Not r Is Nothing Then Map1.Extent = r
```

This line checks to see whether or not a Rectangle object has been returned from the TrackRectangle method.

If it has, the `Extent` property of the map is set to the new Rectangle. Recall that updating a map's Extent property will cause the map to automatically redraw. After this line, the Else block would be skipped and the procedure would exit, resulting in a map of the area specified by the box the user defined.

The block of code after the Else statement occurs when a mouse button other than the number 1 (left) mouse button was used to perform the mouse click. If so, a zoom-out is performed. To perform the zoom-out, the variable `r` is first set to the Rectangle object representing the current Extent of `Map1`, as follows:

```
Set r = Map1.Extent
```

The Rectangle is then increased in size by a factor of 1.5 using the ScaleRectangle method of the Rectangle object:

```
r.ScaleRectangle 1.5
```

The newly scaled Rectangle is then applied to the Extent property of the map in the last line of code in the procedure. This forces the map to redraw. The procedure is exited and the map displays the new map extent.

## The Command2 Button: Using the Default Symbol

The Command2 button (labeled Single Symbol Map in the user interface) will when pressed draw, as shown in the following illustration, all features in the Counties layer (all the county polygons) with a single color. Because the Color property of the Symbol associated with the Counties layer was set to a pale yellow color in the

Form Load procedure, the result is a map with all counties shaded pale yellow.

*The result of clicking the Command2 (Single Symbol Map) button at run time.*

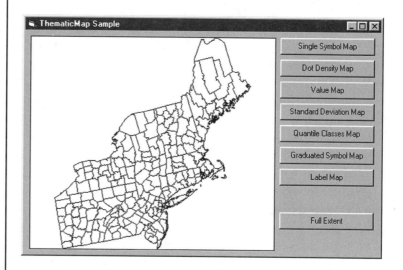

The program code to accomplish this is contained in the Command2 button's Click procedure. When the Command2 button is pressed, the button's Click event occurs. When the Click event occurs, the Command2 Click procedure runs the following program code:

```
Private Sub Command2_Click()
  Map1.Layers("NeCenter").Visible = False  ' hide NeCenter

  Set Map1.Layers("Counties").Renderer = Nothing
  Map1.Refresh
End Sub
```

The first line of code in this procedure turns off the layer named `NeCenter` by setting its `Visible` property to `False`. The next line removes any `Renderer` objects associated with the `Counties` layer by setting the layer's `Renderer` property to the Visual Basic keyword `Nothing`. MapObjects Renderer objects can be associated with a layer by assigning the specific Renderer object to the Renderer property of the layer. When the Renderer property is set to Nothing, as in the code above, any existing reference to a Renderer is removed.

When no Renderer is assigned to a layer, the layer is drawn using its associated Symbol object. The last line of code is:

```
Map1.Refresh
```

This refreshes (redraws) the map. The map is redrawn and the Counties layer is displayed with the entire layer (all county polygons) drawn in the pale yellow color according to the properties of the Symbol associated with the layer.

## The Command1 Button: Using the DotDensityRenderer

The Command1 button (labeled Dot Density Map in the user interface) will when pressed create a dot density map showing population density for the Counties layer. A Command1 button result is shown in the following illustration.

*The result of clicking the Command1 (Dot Density Map) button at run time.*

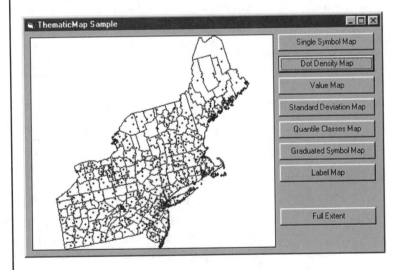

The program code to create this map is contained in the Command1 button's Click procedure. When the Command1 button is pressed, the button's Click event occurs. When the Click event occurs, the following program code is executed within the Command1 Click event procedure:

```
Private Sub Command1_Click()
  Map1.Layers("NeCenter").Visible = False   ' hide NeCenter

  Set ly = Map1.Layers("Counties")
  Set ly.Renderer = New DotDensityRenderer
  ly.Renderer.Field = "HBEDS_1000"

  ' set a dot value
  Set stats = ly.Records.CalculateStatistics("HBEDS_1000")
  ly.Renderer.DotValue = (stats.Min + (stats.Max - stats.Min) / 2) / 20
  Map1.Refresh
End Sub
```

The first line of the code turns off the layer named NeCenter because it will not be used for this map. The second block of code begins with the line:

```
Set ly = Map1.Layers("Counties")
```

This line finds the Counties layer in the Layers collection of Map1 and sets a variable named ly to reference it. This is done so that you can easily refer to the layer later in this code simply by using the variable ly.

The next two lines, which follow, start the setup of the DotDensityRenderer you will use for symbolizing the Counties layer. The first of these two lines creates the DotDensityRenderer.

```
Set ly.Renderer = New DotDensityRenderer
  ly.Renderer.Field = "HBEDS_1000"
```

Note that in this case the DotDensityRenderer is never assigned to a variable. Instead, when it is created using the keyword New, it is immediately associated with the layer by assigning it to the Renderer property of the layer.

The next line assigns a value to the Field property of the DotDensityRenderer that was just assigned to the layer's Renderer property. The Field property references a field in the layer's attribute table, the value of which will be used for creating the dot density patterns. In this case, you use the field named HBEDS_1000, which represents hospital beds per 1,000 people.

The final block of code in this procedure tells the Dot-DensityRenderer how many dots to draw for the values it finds in the HBEDS_1000 field of the Counties layer. This is done using the DotValue property of the DotDensityRenderer. The number of dots drawn for a given polygon will equal the value found in that polygon's attribute table record (in the field set in the Renderer's Field property) divided by the number set in the Dot-Value property.

For example, if the value of HBEDS_1000 is 200 for a given polygon and the DotValue property is set to 50, four dots (200/50) will be drawn. If the DotValue property is set to 100, two dots (200/100) will be drawn.

The first line in the final block of this procedure, which follows, sets a variable named stats to the results of the CalculateStatistics method.

```
Set stats = ly.Records.CalculateStatistics("HBEDS_1000")
```

This method can be applied to a set of table records to calculate statistics for a specified field. These calculations include the minimum value of the set of records, the maximum value, the mean, the sum, and the standard deviation. The CalculateStatistics method returns a Statistics object. The Statistics object stores the calculated results in its properties: Max (maximum value), Min (minimum value), Mean (mean), Sum (sum), and StdDev (standard deviation).

Note that in the previous line of code the `CalculateStatistics` method is applied to the records returned by `ly.Records`. Remember that `ly` refers to the Counties layer; therefore, the `Records` property of `ly` contains the records of the Counties layer attribute table.

The field name for which the calculations will be performed is specified in the parentheses following the call to `CalculateStatistics`, which in this case is `HBEDS_1000`. The result of applying `CalculateStatistics` in this way is a Statistics object containing, in its properties, statistics for the `HBEDS_1000` field.

The final two lines in this procedure complete the setup of the DotDensityRenderer and then redraw the map. These lines are:

```
ly.Renderer.DotValue = (stats.Min + (stats.Max - stats.Min) / 2) / 20
Map1.Refresh
```

The first of these two lines sets the `DotValue` property of the `Renderer`. `DotValue` is set to the result of an equation using the properties of the Statistics object. The equation first determines the midpoint value of the range of values: (minimum value + (maximum value – minimum value ) / 2). This midpoint value is then divided by a value of, in this case, 20.

Increasing this last value of the equation (currently 20) will result in more dots being displayed on the map. This occurs because increasing this value causes the equation to return a smaller number for the `DotValue`. Because the number of dots displayed in a polygon is the value from the specified field of that polygon's attribute record divided by the `DotValue`, a larger number of dots is displayed for that polygon. Decreasing the last value of the equation will have the opposite effect: fewer dots will be displayed on the map.

The final line of code in the procedure invokes the Refresh method of the Map object to redraw the map using the DotDensityRenderer defined. The result is a dot density map displaying population density for the counties in the Counties layer.

## The Command3 Button: Using the ValueMapRenderer

The Command3 button (labeled Value Map in the user interface) will when pressed create a map, such as that shown in the following illustration, showing each county in the Counties layer displayed with a color identifying the state in which the county is contained. For example, all counties in the state of New York may be colored blue, whereas all counties in the state of Maine may be colored red. The result is a county map that displays states (by color) as well as counties (by boundary lines).

*The result of clicking the Command3 (Value Map) button at run time.*

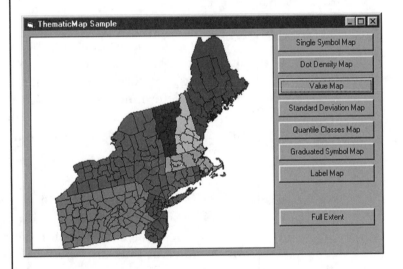

To do this, the ValueMapRenderer object is used. The program code to create this map is contained in the Command3 button's Click procedure. When the Command3 button is pressed, the button's Click event occurs. When the Click event occurs, the following program code is executed within the Command3 Click event procedure:

```
Private Sub Command3_Click()
  Map1.Layers("NeCenter").Visible = False   ' hide NeCenter

  ' find unique values for STATE_NAME field
  Dim strings As New MapObjects.strings
  Set ly = Map1.Layers("Counties")
  Set recs = ly.Records
  Do While Not recs.EOF
    strings.Add recs("STATE_NAME").Value
    recs.MoveNext
  Loop

  Set ly.Renderer = New ValueMapRenderer
  ly.Renderer.Field = "STATE_NAME"

  ' add the unique values to the renderer
  ly.Renderer.ValueCount = strings.Count
  For i = 0 To strings.Count - 1
    ly.Renderer.Value(i) = strings(i)
  Next i

  Map1.Refresh

End Sub
```

The first line of code in this procedure turns off display of the layer named NeCenter because this layer is not required for this procedure. The first block of code in the procedure finds and stores a list of unique values found in the Counties layer. The first line of this block, which follows, declares and creates a new Strings object named strings.

```
Dim strings As New MapObjects.strings
```

The Strings object in MapObjects is a collection to which you can add strings. By default, a Strings object will contain only unique values. For instance, if you have previously added the string "apple" to a Strings collection object, a subsequent attempt to add the identical string, "apple," will not result in two "apple" strings in the collection. The Strings object will ignore the second attempt because it will recognize that it would result in non-unique, duplicate storage of the string "apple." The next two lines in the first block of code are:

```
Set ly = Map1.Layers("Counties")
 Set recs = ly.Records
```

These lines set a variable named ly to reference the Counties layer of the map named Map1, and set a variable named recs to reference the attribute table records of ly (the Counties layer). Referencing the Records property of a layer, as done here, returns a Recordset object that refers to the set of records contained in the layer's attribute table. The following Do loop ends the first block of code in this procedure:

```
Do While Not recs.EOF
    strings.Add recs("STATE_NAME").Value
    recs.MoveNext
Loop
```

The line beginning the Do loop is:

```
Do While Not recs.EOF
```

This line specifies that this loop will continue to process as long as (While) you are not at the end of file (EOF), meaning the end of the Recordset.

This means that the code in the Do loop will execute once for every record in the Recordset. Therefore, the code will execute for every record in the Counties layer attribute table. The first line in the body of the loop is:

```
strings.Add recs("STATE_NAME").Value
```

This line retrieves the value in the STATE_NAME field of the current record in the Recordset by using the Value property.

When the loop first begins execution, it finds that the current record is the first record of the Recordset. The Add method of the Strings object is invoked to add the returned value of STATE_NAME to the Strings object named `strings`. Remember that the value will only be added if it is unique. Therefore, this code creates a list of strings representing all unique values for the STATE_NAME field of the Counties layer attribute table.

The last of the two lines found in the body of the Do loop invokes the MoveNext method of the Recordset object to point to the next record of the recordset, as follows:

```
recs.MoveNext
```

The end of the Do loop is then reached and the Loop statement sends you back to the top of the loop. The body of the loop is then processed for the next record (to which you were moved by the previously invoked MoveNext method).

This process continues until MoveNext eventually places you at EOF (the end of the recordset) and the loop processing terminates. After the EOF is reached and the Do loop terminates, the Strings object named strings will contain a complete list of all unique values for the STATE_NAME field. This list of unique values is a list of all states represented by at least one county in the Counties layer. The second block of code in the Command3 button's Click procedure consists of the following two lines of code:

```
Set ly.Renderer = New ValueMapRenderer
ly.Renderer.Field = "STATE_NAME"
```

The first of these lines creates a ValueMapRenderer object and immediately assigns it to the Renderer property of the layer ly (the Counties layer). The second line then sets the Field property of the new renderer to

STATE_NAME. The `Field` property of the `ValueMa-pRenderer` tells the `ValueMapRenderer` which field contains values that will determine how a feature will be rendered. This is the same field from which you created the list of unique values: STATE_NAME.

The third block of program code in this procedure completes the setup of the ValueMapRenderer, as follows:

```
' add the unique values to the renderer
  ly.Renderer.ValueCount = strings.Count
  For i = 0 To strings.Count - 1
    ly.Renderer.Value(i) = strings(i)
  Next i
```

The first line retrieves the number of unique values stored in strings by invoking the `Count` method of the Strings object. This number is then assigned to the `ValueCount` property of the ValueMapRenderer associated with `ly` (the Counties layer). The `ValueCount` property represents the number of unique values the ValueMapRenderer will be expected to render.

The following line begins the For loop. This line specifies that the program code contained within the loop will process once for each string in `strings`.

```
For i = 0 To strings.Count - 1
```

This is accomplished by initializing to 0 a counter variable named `i`. This variable will be incremented by 1 each time the `For` loop executes. The loop will stop execution when `i` is incremented beyond the value of `strings.Count - 1`. The range of values the counter will assume, from 0 to `strings.Count - 1`, will cover the index values for every String object contained in the Strings collection (`strings`) because String objects contained in a Strings object are indexed beginning at 0.

The following line constitutes the body of the For loop. This line sets the Value property of the ValueMapRenderer.

```
ly.Renderer.Value(i) = strings(i)
```

The `Value` property is an array containing the unique values found in the field specified in the ValueMapRenderer's Field property (in this case, STATE_NAME). The counter `i` is used as an array subscript to reference specific elements in the `Value` array and the `strings` String collection. On the first pass through this loop, this statement will be interpreted as the following:

```
ly.Renderer.Value(0) = strings(0)
```

On the second pass, it will be interpreted as the following:

```
ly.Renderer.Value(1) = strings(1)
```

and so on.

The result is that each of the unique values in the Strings collection named `strings` (recall that these are the unique values of the STATE_NAME field) are loaded into the `Value` array. The `Next` statement in the last line of this loop increments the counter, `i`, and returns to the For loop, where `i` is evaluated to see if it is still within the range allowing execution of the loop. To summarize the programming for the Command3 button, you have:

1. Created a ValueMapRenderer.

2. Assigned the ValueMapRenderer to a layer.

3. Set the ValueMapRenderer's Field property to tell the ValueMapRenderer which field of the feature attribute table will have its value compared to the ValueMapRenderer's Value array for determining how to symbolize the feature. Note that a Symbol array property of the

ValueMapRenderer can be used to assign unique symbols for each value in the Value array. For example, the symbol of Value(1) would be Symbol(1). Alternatively, you can use the default symbols, as done in the programming of the Command3 button.

4.  Set the ValueMapRenderer's ValueCount property to tell the ValueMapRenderer how many items are in the Value array.

5.  Set the ValueMapRenderer's Value array to contain a list of unique values that will be found in the field referenced by the Field property.

All that remains is to redraw the layer. The Refresh in the final line of this procedure redraws Map1, and therefore the layer with the newly applied ValueMapRenderer. The result is a map showing each county in the Counties layer displayed with a color identifying the state in which the county is contained.

## The Command8 Button: Using the ClassBreaksRenderer

The Command8 button (labeled Quantile Classes Map in the user interface) will when pressed create and map several classes, each representing a specific range of population values that will contain the same quantity of counties as any other class (hence the name Quantile). The population values used in this case are from the field named P_OTHER, which represents the percentage of the population that is of unidentified ethnic origin. The map will display each county in the Counties layer in a color identifying the class in which the county is contained, as shown in the following illustration.

*The result of clicking the Command8 (Quantile Classes Map) button at run time.*

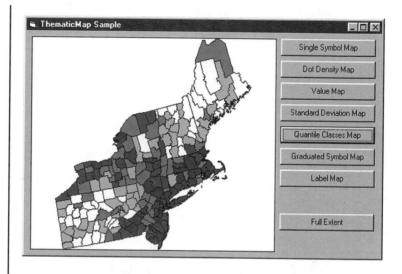

The ClassBreaksRenderer is used to create a map of this type. Notice the similarity of this type of map to the type produced by the ValueMapRenderer. The difference is that the ValueMapRenderer represents unique values, whereas the ClassBreaksRenderer represents ranges of values, called classes.

Before proceeding, note that the Standard Deviation Map and Graduated Symbol Map buttons also use the ClassBreaksRenderer to produce their more sophisticated variations of class breaks. Because it provides a simple introduction to the ClassBreaksRenderer, the Quantile Classes Map will be reviewed in this section, with the Standard Deviation Map and Graduated Symbol Map buttons left for your further investigation.

The program code to create the Quantile Classes Map is contained in the Command8 button's Click procedure. When the Command8 button is pressed, the button's Click event occurs. When the Click event occurs, the following program code is executed within the Command8 Click event procedure:

```
Private Sub Command8_Click()
  Map1.Layers("NeCenter").Visible = False  ' hide NeCenter

  Set ly = Map1.Layers("Counties")
  Set ly.Renderer = New ClassBreaksRenderer
  Set r = ly.Renderer

  nClasses = 5
  nRecs = ly.Records.Count
  r.BreakCount = nClasses - 1
  r.Field = "P_OTHER"

  ' query all the features and order the results
  Set recs = ly.SearchExpression("FeatureId > -1 order by P_OTHER")

  ' navigate the record set and set up the breaks
  For i = 0 To r.BreakCount - 1
    For j = 1 To nRecs / nClasses
      recs.MoveNext
    Next j
    r.Break(i) = recs("P_OTHER").Value
  Next i

  ' create a color ramp
  r.RampColors moLightYellow, moBlue
  Map1.Refresh

End Sub
```

The first line of the code turns off the layer named NeCenter because NeCenter will not be used to create the quantile classes map. The second block of code begins with the following line:

```
Set ly = Map1.Layers("Counties")
```

As in the DotDensityRenderer example reviewed earlier, this block sets a variable named ly to reference a MapLayer, and then creates the Renderer and assigns it to the layer's Renderer property. Note that a ClassBreaksRenderer is the type of Renderer created for this example, and that a variable named `r` is created to reference the Renderer object. The next block in this procedure contains the following program code:

```
nClasses = 5
nRecs = ly.Records.Count
r.BreakCount = nClasses - 1
r.Field = "P_OTHER"
```

The first two lines initialize two variables that will be used later. The first, `nClasses`, is set to indicate the number of classes you will create, which in this case is 5. The second variable, `nRecs`, is set to indicate the number of records found in the Recordset object (the attribute table) of layer `ly` (the Counties layer).

The last two lines of this block set two properties of the ClassBreaksRenderer: `BreakCount` and `Field`. `BreakCount` indicates the number of breaks (divisions between classes) the ClassBreaksRenderer will need to deal with. This will be one less than the number of classes, as represented by `nClasses` − 1 in the line of program code.

The `Field` property specifies the name of the field that will be referenced to determine what class a record is contained in. In this case, you assign the field name P_OTHER to this property. The next line of program code is:

```
Set recs = ly.SearchExpression("FeatureId > -1 order by P_OTHER")
```

This line creates a Recordset object (named `recs`) containing all records with valid values in the field named `FeatureId`, sorted in ascending order by the values in the field named P_OTHER. This is done using the `SearchExpression` method of the MapLayer object.

SearchExpression is a method that will query a layer attribute table using a Standard Query Language (SQL) expression you supply. SQL is a standardized language for querying databases. Although most database vendors have their own specific extensions to SQL to provide additional query capabilities, most major database vendors support the core SQL standard. This means that a standard SQL query should work with any database supporting standard SQL.

In the previous line of program code, the SQL expression is contained in the parentheses following SearchExpression. Notice that the SearchExpression method is invoked for the layer ly (the Counties layer). The expression supplied searches the layer's attribute table for all records containing a value greater than –1 in the field named FeatureId. The resulting set of valid records is then sorted by the values found in the field named P_OTHER. This is done using the SQL order by statement. The next block of program code in the Command8 Click event procedure contains the following lines:

```
For i = 0 To r.BreakCount - 1
    For j = 1 To nRecs / nClasses
      recs.MoveNext
    Next j
    r.Break(i) = recs("P_OTHER").Value
Next i
```

Note that this block contains a nested For loop, which is a For loop contained within another For loop. The line beginning the outer For loop is:

```
For i = 0 To r.BreakCount - 1
```

This line specifies that the program code contained within the outer loop will execute the same number of times as the value of BreakCount, which indicates the number of breaks between classes, which in this case is four.

The effect of this For loop is to execute the code inside the loop once for each class break, with the class break represented by the value in the counter, i. The counter i represents the first break when its value is 0, the second break when its value is 1, and so on.

The first statement encountered in the outer For loop is another For loop. The first line in this inner For loop is:

```
For j = 1 To nRecs / nClasses
```

This line specifies that the program code contained within the inner loop will process once for each value in the range 1 to the number of records (nRecs) divided by the number of classes (nClasses).

For instance, if the number of records in the Counties layer were 20, and the number of classes to be defined were 5, this inner For loop would execute 20/5 (4) times. This is significant because the only line of code contained within this loop is a MoveNext statement that points you to the next record in the Recordset. This means that when the inner For loop is exited, you will be positioned at a record that is on the boundary of a break between classes.

After the inner For loop is exited, and you are left pointed to a record that represents a class break, you proceed to the next line of code in the outer For loop, which follows. This line of program code sets the Break property of the ClassBreaksRenderer.

```
r.Break(i) = recs("P_OTHER").Value
```

The Break property is an array of values that indicates the upper boundary values of a class. Because you set the Renderer to have a BreakCount of 4, the Renderer will have four values in the Break array, each representing the upper boundary value of the class range ended by the corresponding break.

Recall that the Recordset referred to by recs is sorted in ascending order of the values in P_OTHER. Also recall that the inner For loop previously executed has left you pointed to a record representing a class break. Because of this state of affairs, setting the value of the Break array to the value of the current record tells the Renderer the proper P_ORDER value at which the class break should be defined.

The outer and inner For loops continue to process until all four breaks have corresponding upper boundary values stored in the Break array property of the renderer. All that remains to be done is to assign colors to the ClassBreaksRenderer classes. This is accomplished by the remaining lines of program code in this procedure, which follow.

```
r.RampColors moLightYellow, moBlue
Map1.Refresh
```

The first line of code invokes the RampColors method of the ClassBreaksRenderer object referred to by the variable r. The RampColors method assigns a color to the first and last categories of a ClassBreaksRenderer object and then interpolates the color for each intervening category.

In this code, you are specifying that the first class must be colored light yellow, the last class colored blue. MapObjects is left to define a graduated range of colors between light yellow and blue and then assign the colors to the classes. To summarize the programming for the Command8 button, you have:

1. Created a ClassBreaksRenderer

2. Assigned the ClassBreaksRenderer to a layer

3. Set the ClassBreaksRenderer BreakCount property to tell the ClassBreaksRenderer the number of divisions between classes

4. Set the ClassBreaksRenderer Field property to specify the field whose value will determine what class a particular record should be assigned to

5. Set the ClassBreaksRenderer Break property to contain a list of values determining the upper boundary of each class

**6.** Assigned a graduated set of colors to each of the Class-BreaksRenderer's classes

All that remains is to invoke the Refresh method in the last line of this procedure to redraw the map with its newly applied ClassBreaksRenderer. The result is a map of counties with each county color-coded to show its assigned class, with each class range containing the same number of counties.

## The Command6 Button: Using the LabelRenderer

The Command6 button (labeled Label Map in the user interface) when pressed creates labels for the polygon features in the Counties map layer, as shown in the following illustration. The map displays each county in the Counties layer with a label indicating the county name. The LabelRenderer object is used to create a map of this type. The text displayed is taken from the values of a specified field in the layer's attribute table.

*The result of clicking the Command6 (Label Map) button at run time.*

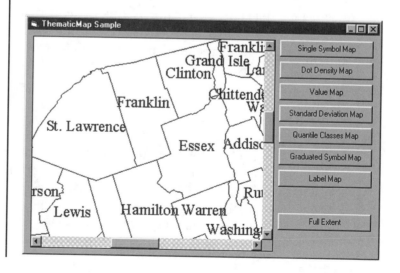

The program code to create the labeled map is contained in the Command6 button's Click procedure. When the Command6 button is pressed, the button's Click event occurs. When the Click event occurs, the following program code is executed within the Command6 Click event procedure:

```
Private Sub Command6_Click()
  Map1.Layers("NeCenter").Visible = False   ' hide NeCenter

  Dim f As New StdFont
  f.Name = "Times"
  f.Bold = False

  Set ly = Map1.Layers("Counties")
  Set ly.Renderer = New LabelRenderer
  ly.Renderer.Symbol(0).Height = 12000
  Set ly.Renderer.Symbol(0).Font = f
  ly.Renderer.Field = "cnty_name"
  ly.Renderer.AllowDuplicates = True
  Map1.Refresh
End Sub
```

The first line of the code turns off the layer named NeCenter because NeCenter will not be used to create the labeled map. The first block of code in this procedure is:

```
Dim f As New StdFont
  f.Name = "Times"
  f.Bold = False
```

This block creates a Font object to be used when labeling the map. A Font object is provided by Visual Basic to contain information needed to format text for display. Objects that display text generally provide a Font property to which a Font object can be assigned. In MapObjects, the Symbol object and the TextSymbol object support the Font property.

To create a Font object in Visual Basic, the StdFont object is used. After it is created, the Font object's properties may be modified as required. Then the Font object can be applied to the Font property of the object that will display the text—a MapObjects Symbol object, for example. The object displaying text will use the Font object and its properties to determine how the text should appear. The first line of this block of code creates a Font object named f, as follows:

```
Dim f As New StdFont
```

The next two lines set two of the properties of the Font object named f. The Name property is used to identify the name of the character set the font will use.

```
f.Name = "Times"
f.Bold = False
```

In this case, you specify the Times character set. The Bold property specifies whether text drawn using this font should be drawn bold. In this case, you set the Bold property to False to indicate that text drawn with this font should *not* be drawn bold.

The last block of code in this procedure is where the LabelRenderer is constructed and applied. The first two lines of this block set a variable named ly to reference the Counties layer of the map named Map1, then create a LabelRenderer and assign it to the layer's Renderer property as follows:

```
Set ly = Map1.Layers("Counties")
Set ly.Renderer = New LabelRenderer
```

The last few lines of code in this procedure set properties of the LabelRenderer now assigned to the layer ly (the Counties layer), and then redraw the map, as follows:

```
ly.Renderer.Symbol(0).Height = 12000
Set ly.Renderer.Symbol(0).Font = f
ly.Renderer.Field = "cnty_name"
ly.Renderer.AllowDuplicates = True
Map1.Refresh
```

The first of these lines sets the `Height` property of the LabelRenderer's Symbol object.

The Symbols associated with a LabelRenderer are TextSymbol objects. A LabelRenderer may have many TextSymbol objects associated with it. The Symbol property is therefore an array of TextSymbol objects.

Although not used in this example, the SymbolField property of the LabelRenderer can be used to specify a field whose value indicates which TextSymbol in the Symbol array should be used for a given record. In the example code shown here, you reference only Symbol(0), the first Symbol in the Symbol array, and do not use the SymbolField property because you are using only one TextSymbol for all text drawn with this LabelRenderer.

The `Height` property of the TextSymbol object is used to specify the height, in map units (e.g., feet), of the text when drawn. This value does not vary once you set it; therefore, setting this property has the effect of producing scale-dependent text, as shown in the following illustration.

*The effect of scale-dependent text. Notice the increasing size of the text as the map is zoomed in.*

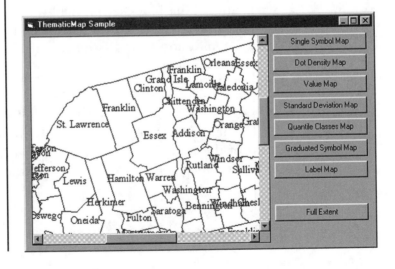

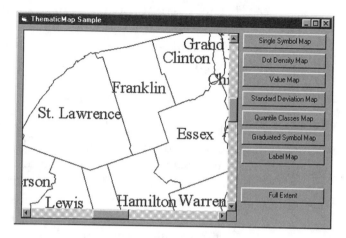

For instance, if you set this property to 10, and your map units are in feet, text drawn on the map will always be 10 feet in height relative to other features on the map. Zooming out will cause 10 feet to appear smaller on the map display, and therefore the text will appear smaller. Zooming in will have the opposite effect. In the case of this example, you set the Height property to 12000.

The next line of code assigns the previously created Font object, f, to the Font property of the LabelRenderer's TextSymbol object, Symbol(0), as follows:

```
Set ly.Renderer.Symbol(0).Font = f
```

Any text now drawn with this LabelRenderer will use this font for determining how to display the text. Notice that you are using a combination of TextSymbol properties (such as Height) and Font object properties (such as Bold) to define how the text should look. The next two lines of program code in this block set the Field and AllowDuplicates properties of the LabelRenderer, as follows:

```
ly.Renderer.Field = "cnty_name"
ly.Renderer.AllowDuplicates = True
```

The `Field` property identifies which field will be used as the source for the label text. Attribute values from this field will provide text for the features associated with each record. In this example, the `Field` property is set to `cnty_name` to identify the field containing county names as the source for the label text.

The `AllowDuplicates` property specifies whether a LabelRenderer will draw a duplicate label if it has already drawn a label with the same text. In working with street data, for instance, it may be redundant to allow duplicates because a street with many individual segments will display the same text repeatedly. In this example, however, you will allow duplicate labels to be drawn.

The following illustration shows the effect of an AllowDuplicates property set to True. Note the repeating county name Jefferson. The label appears once for each polygon representing a portion of Jefferson county.

*The effect of setting the AllowDuplicates property to True.*

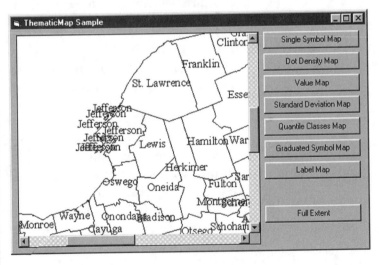

The following illustration shows the same map with the AllowDuplicates property set to False. Note that the county name Jefferson no longer repeats. The label appears only once on the map, regardless of the number of polygons representing Jefferson county.

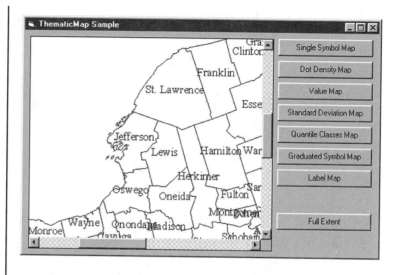

*The effect of the setting the AllowDuplicates property to False.*

The Refresh method is invoked in the last line of this procedure to redraw the map with its newly applied Label-Renderer. The result is a map displaying each county in the Counties layer with the county's name as a label.

# The Command4 and Command5 Buttons: Further Uses of the ClassBreaksRenderer

As noted in the discussion of the ClassBreaksRenderer in the section on the Command8 button procedure, the Standard Deviation Map (Command4) button and the Graduated Symbol Map (Command5) button use the ClassBreaksRenderer to produce their more sophisticated variations of class breaks.

The Standard Deviation Map button when pressed will shade the counties in the Counties layer according to a calculation of standard deviation, as shown in the following illustration.

*The result of clicking the Command4 (Standard Deviation Map) button at run time.*

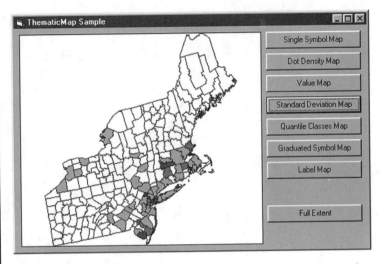

The Graduated Symbol Map button uses the same calculation of standard deviation to display symbols of increasing or decreasing size, depending on the value, as shown in the following illustration. Although these two example commands are not reviewed in this book, you are encouraged to review them on your own as further examples of the flexibility of the Renderer objects.

*The result of clicking the Command5 (Graduated Symbol Map) button at run time.*

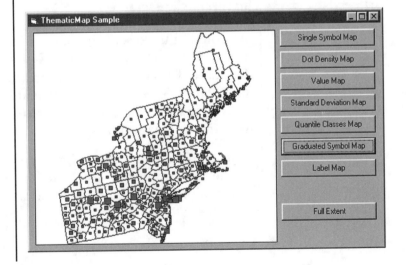

# 9

# Analyzing Maps

## MapObjects Recordsets, Geometric Objects, and Spatial Queries

GIS incorporates geography into data analysis.

As discussed in Chapter 2, easily creating and displaying sophisticated maps is only part of the power of GIS technology. In addition to creating and displaying maps, geographic technologies also provide the ability to analyze information in a way that incorporates the geographic aspect of data into an analysis.

Geography incorporation requires integrated access to spatial and aspatial data.

Incorporating the geographic aspect of data into analysis requires integrating both spatial (geographic) and aspatial (nongeographic, or attribute) information during data browsing and query. It also requires the provision of tools to perform the sort of specialized analysis (spatial analysis) possible using the geometry of mapped data. This chapter examines the support provided by MapObjects for this type of integrated data query and analysis.

## Multi-dimensional Query and Analysis

One of the unique contributions of GIS to the general information technology industry has been the stimulation of an awareness of the importance of spatial data. Useful analysis has been, and will continue to be, possible with aspatial data alone. However, when spatial data is combined with aspatial data, there are new possibilities for data analysis.

Consider the case of a retailer needing to summarize information regarding customers within a specific trade area, living three miles from a major transportation corridor, over 65 years of age, and with an income over $50,000. A query of this type involves both a spatial dimension (*points* representing customers are *located* within a given *polygon* representing a trade area and are *proximal* to transportation corridors) and an aspatial dimension (customer *records* in the *table* resulting from the query have a value in the age *field* that is greater than 65 and a value in the income *field* that is greater than $50,000).

*An example of a complex query using spatial and aspatial data.*

Geographic technology typically provides a means of viewing data from two dimensions: spatial and aspatial. A *spatial* dimension usually involves geometric shapes, and locations of those shapes, within coordinate systems. An *aspatial* dimension usually involves tables with columns, rows, and values.

*Spatial* and *aspatial* refer to two dimensions of the same data.

A data set may exist as a single data set, such as the customer database in the previous example, but the data may be viewed from either spatial or aspatial aspects. In most well-designed GIS systems, either of these dimensions provides an access point for reaching the other dimension of the data. The integration of spatial and aspatial analysis is shown in the following illustration.

*Multidimensional queries can integrate spatial and aspatial analysis.*

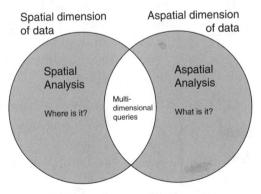

It should be noted that some database management systems are beginning to support other dimensions of data, such as the temporal dimension, involving location in time. Along with support for storage of these additional data dimensions, tools are being developed to support query of them. This will allow incorporation of many other dimensions (such as temporal) of data into complex queries along with the spatial and aspatial dimensions previously referred to.

MapObjects provides Recordset, Table, and associated objects as a means of approaching the aspatial dimension of data, and Geometric objects and methods as a means of approaching the spatial dimension of data. Properties and methods of these various objects, along with query methods associated with MapLayer objects, enable the integration of these two dimensions in query and analysis. The remaining sections in this chapter examine these objects, properties, and methods.

*DBMSs now support additional dimensions, such as time.*

*MapObjects supports spatial and aspatial dimensions.*

# The Recordset Object

The sections that follow explain what a Recordset object is, how to access the Recordset of a MapLayer object, and inspecting and updating Recordsets. The topic of inspection and update includes navigating in a Recordset, and modifying fields in a Recordset.

## What Is a Recordset?

*Visual Basic and MapObjects both provide Recordset objects.*

Readers with experience in Visual Basic will already be aware of the concept of a Recordset object. Visual Basic provides a Recordset object to manipulate data in a database at the record level. The object itself represents records in a base table or in the set of records resulting from a query. MapObjects also provides a Recordset object for manipulating record-based data.

The MapObjects Recordset object is similar to the Visual Basic Recordset object in that it represents the records in a table of data or the records resulting from a query. Because of this similar purpose, the MapObjects Record-

*MapObjects' Recordset object is similar to Visual Basic's.*

set also provides many methods and properties similar to the Visual Basic Recordset object, such as the MoveNext method to move to the next record in a set, and the Updatable property to indicate whether changes can be made to the records in a Recordset object.

*MapObjects Recordsets are derived from GeoDataset objects.*

The MapObjects Recordset object differs from the Visual Basic Recordset object primarily in regard to its source and method of creation. In Visual Basic, Recordset objects are created by direct references to tables either by table name or through a query. In MapObjects, Recordset objects are created by indirectly referencing table data through a GeoDataset object.

## Accessing the Recordset of a MapLayer Object

When a MapLayer object is created, a Recordset object is also created by MapObjects to represent the records of the GeoDataset object associated with that MapLayer object. This Recordset object is accessed via the Records property of the MapLayer object. Referencing this property returns the Recordset representing the full set of records of the MapLayer's data.

*GeoDataset objects can incorporate aspatial data.*

Recall that a GeoDataset object represents a layer of geographic data and is associated with a MapLayer object. A GeoDataset object has spatial data in the form of georeferenced geometric shapes. Associated with the spatial data, and existing as part of the GeoDataset object, are table records with one record per spatial feature. These records contain aspatial data directly associated with the spatial feature.

*Recordsets can represent all records of a Geodataset or only a subset of the records.*

A Recordset object representing these records of a GeoDataset object can be accessed via the Records property of the MapLayer object with which the GeoDataset is associated. For instance, if you have a MapLayer named TheRoads existing as a layer in a map named Map1 and want to access the records of the layer, you might write the following program code:

```
Dim recset As MapObjects.Recordset
Set recset = Map1.Layers("TheRoads").Records
```

There are several points of interest in this code fragment. First, note how the variable `recset` is declared an object of type *MapObjects*.Recordset. A full reference to the data type is used, including a reference to the MapObjects library. In most cases this full declaration is optional, but should be done as a matter of good form. However, in this case it is necessary to fully qualify the declaration because this ensures that Visual Basic understands that the Recordset object being declared is the MapObjects Recordset object rather than the Visual Basic Recordset object.

> A MapObjects Recordset requires a fully qualified declaration.

Second, note that the declaration does not include the New keyword, as have most of the declarations discussed to this point. This is because a Recordset object is a *dependent* object rather than a *creatable* object. Creatable objects can be instantiated with the New keyword, as in the following code:

> Recordsets are dependent objects.

```
Dim ml As New MapObjects.MapLayer
```

Dependent objects are lower in the class hierarchy and as such can only be accessed by using a property or method of a creatable object to return a reference to the dependent object. This is done in the previous code example by using the Records property of the MapLayer object (TheRoads) to return the Recordset object:

> Dependent objects are referenced through methods or properties.

```
Set recset = Map1.Layers("TheRoads").Records
```

In other words, some objects (creatable) can exist on their own, whereas other objects (dependent) cannot exist independent of another type of object. Specifically, in the case under discussion, a MapObjects Recordset object cannot exist independently of a GeoDataset object associated with a MapLayer object.

The following illustration shows a class hierarchy diagram depicting the concepts of creatable and dependent

Dependent objects are
lower in the class hierarchy.

objects. In this diagram, a MapLayer object is shown higher in the class hierarchy. The MapLayer object can exist independent of other objects and can be created by a client. It is therefore a creatable (independent) object.

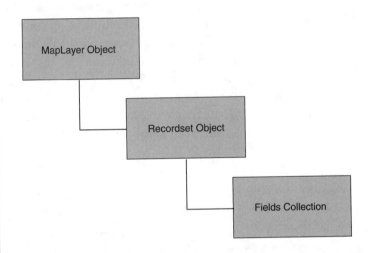

*A class hierarchy diagram showing creatable and dependent objects.*

A Recordset object requires that a MapLayer exist. The MapLayer is created by the MapLayer object when the MapLayer object is created by a client. It can only be accessed by referring to its associated MapLayer object. It is therefore a dependent object. Still lower in the class hierarchy—a Fields collection containing all fields of a recordset—exists only in association with a Recordset object. It can only be accessed by referring to the associated Recordset object. Therefore, it is also a dependent object.

Returning to the previous code fragment, also notice that the Visual Basic keyword Set is used to assign an object reference to the variable. With creatable objects, this assignment is not necessary because the object is instantiated with the New keyword, often in the same line as the declaration made with the Dim statement. With dependent objects, however, the Dim statement merely declares the variable. The Set statement is used to instantiate the object by setting a reference to the dependent object when it is returned by the method or property.

The Set keyword is used
to reference
dependent objects.

## Inspecting and Updating Recordsets

The Recordset object provides properties and methods that can be used to inspect and modify the records contained in the Recordset object. The following sections discuss navigating records, editing records, and modifying fields in a Recordset.

### Navigating Records in a Recordset

Several Move methods allow record positioning.

The Recordset object provides the following methods to reach a particular position in the records of a Recordset object:

❑ *MoveFirst* to move to the first record

❑ *MovePrevious* to move to the record previous to the current record

❑ *MoveNext* to move to the record subsequent to the current record

In conjunction with the EOF (end of file) property, which indicates when the end of records has been reached, these methods can be used to iterate through all records in a Recordset object. For instance, to move one at a time through all the records of the Recordset associated with a layer named Roads, you might write the following code:

```
Dim recSet as MapObjects. RecordSet
Set recSet = Map1.Layers("Roads").Records
Do While Not recSet.EOF
              [The program does something with the record here….]
              recSet.MoveNext
Loop
```

### Editing Records in a Recordset

A Recordset object is not necessarily updatable. Operating system permissions on a shapefile, for example, may make the shapefile read-only. To determine whether the Recordset object is updatable, you may query the Updat-

The Updatable property indicates editing permission.

able property of the Recordset object. The Updatable property returns a Boolean (True or False) value indicating whether or not records in the Recordset object can be edited.

If a Recordset object is updatable, the Edit method may be invoked to begin editing on the current record. Changes made can then be saved by invoking the Update method, or canceled by invoking the CancelUpdate method. The editing session can then be ended by invoking the StopEditing method. New records can be added using the AddNew method, or deleted by using the Delete method.

Recordset methods provide editing capabilities.

To access and update the actual values in a Recordset object requires accessing two dependent objects: the Fields collection and the Field object. A Field object represents a column of data in the table represented by the Recordset object. Each field has a name identified in the Name property, a data type stored in the Type property to indicate the type of data stored in the column (e.g., string, long, or double), and a Value property that returns the value stored in the column for the current record.

Field objects provide access to data values.

A Fields collection contains all Field objects of the Recordset object. Field objects can be accessed from a Fields collection by using the Item method of the Field collection and supplying an index number or name—a process similar to retrieving a layer from the Layers collection. Recordset object fields, records, and positioning methods are shown in the following illustration.

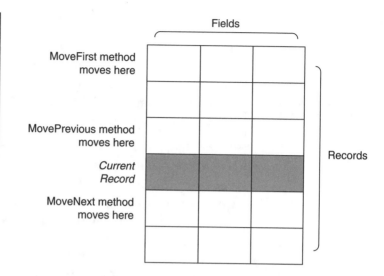

*Recordset object fields, records, and positioning methods.*

Using the previous code example of moving through records of the Recordset object associated with a layer named Roads, you can now insert a meaningful operation to update data values through the use of the Fields collection and the Field object. Suppose the Recordset object of the Roads layer contains two columns, named Miles and Feet. If a value exists in the Miles column, a value can be calculated for the Feet column of every record by modifying the previous code as follows:

*An editing example.*

```
Dim recSet as MapObjects. RecordSet
Dim milesField as MapObjects.Field
Dim feetField as MapObjects.Field
Set recSet = Map1.Layers("Roads").Records
Set milesField = recSet.Fields("Miles")
Set feetField = recSet.Fields("Feet")
If recSet.Updatable Then
Do While Not recSet.EOF
                recSet.Edit
                Set feetField.Value = milesField.Value * 5280
                recSet.Update
                    recSet.MoveNext
Loop
End If
```

## Modifying the Structure of a Recordset

Use the TableDesc object to modify data structures.

The preceding section examined how the Fields collection and Field object allow the modification of data values. The Recordset object has an associated object to allow modification of the actual data structure of fields. This object is called the TableDesc object.

A TableDesc modification example.

The TableDesc object provides properties that allow you to inspect or modify field characteristics. These include Field-Count for the number of fields in the table, FieldName for the name of the field, and FieldType for the type of data stored in the field. These properties are often used when creating a entirely new Recordset, as when you create a shapefile. For instance, you might write the following code to create a GeoDataset of points that will be named Schools and that will contain two associated fields in its Geo-Dataset's recordset (one for storing the school's name, and another for storing the school's ID number):

```
Dim tDesc As New TableDesc
tDesc.FieldCount = 2   ' define two additional fields
tDesc.fieldName(0) = "SchoolName"  'set the field names
tDesc.fieldName(1) = "SchoolID"
tDesc.FieldType(0) = moString  'set the type of each field
tDesc.FieldType(1) = moLong
tDesc.FieldLength(0) = 30  ' set the length of the fields
tDesc.FieldLength(1) = 15
'create the shapefile by adding the geodataset to a dataconnection named 'dc'
Set gd = dc.AddGeoDataset("MyNewGeoDataset", moPoint, tDesc)
```

# Using the Table Object to Access Additional Tabular Data

The sections that follow explain what a table is, how tables work with Open Database Connectivity (ODBC), and how tables are related to MapLayers. The section on rotating tables to MapLayers provides a step-by-step procedure for linking a Table object and a MapLayer object's Recordset object.

## What Is a Table?

*Data outside the GeoDataset may be related to a MapLayer.*

A Recordset object refers to the table associated with a MapLayer object as part of its GeoDataset object. However, there will often be data that is relevant to the MapLayer but that is stored in database tables not directly associated with the MapLayer or its GeoDataset. For example, the GeoDataset named Schools in the previous example may exist in an environment in which there are also tables in a database that store information about individual schools, such as school population and level.

*Tables allow relevant data to be associated with a MapLayer object via the MapLayer's Recordset object.*

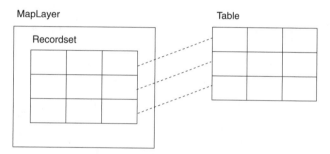

*The Table object represents related data.*

Linking the Schools map and its GeoDataset with the database tables containing school information can enhance opportunities for data analysis. To accomplish this linking, some means must exist for representing the relevant database tables and for creating an association between the database tables and the MapLayer object's Recordset. MapObjects supplies the Table object to represent the database table. A reference to a specific database is made by using the Database and Name properties of the Table object

*Tables can be joined to the relevant Recordset.*

Generally, a Table object is a logical representation of a physical table in a database. A Table object is a read-only data access object, meaning that edits to database tables are not supported by the Table object.

## Tables and ODBC

*ODBC is Open Database Connectivity.*

The Table object can represent a table from any ODBC (Open Database Connectivity) data source. ODBC is a standard protocol for accessing databases. Databases that support ODBC generally provide ODBC software (called drivers). These drivers enable applications requiring database access to speak to the database using ODBC as the "middleware," or negotiator, to work out the details of the access activity.

*ODBC enables DBMS-independent applications.*

The goal of the ODBC approach is to enable the development of DBMS-independent applications, an example of which is shown in the following illustration. An application written to the ODBC specification can speak to any ODBC driver, and therefore any DBMS for which an ODBC driver exists.

*A DMBS-independent application using ODBC.*

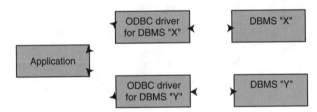

This means that the application is not required to understand the access methods of any particular DBMS. The ODBC driver understands the specific access methods of the DBMS it supports, and handles the translation of application requests into requests understood by the DBMS.

*ODBC data sources relate drivers and data providers.*

An ODBC data source can be established on your computer to indicate what ODBC driver to use when accessing a specific data provider, such as the Oracle RDBMS. Before establishing a relationship between a Table object and a MapLayer's Recordset object, an ODBC data source must be defined for the data provider that contains the data to be represented by the Table. This can be done using utilities of your computer's operating system, such as the ODBC manager located in the Control Panel of Windows 95.

## Relating Tables to MapLayers

*An example of relating a Table to a MapLayer's Recordset.*

Creating a link between a database table represented by a Table object and a MapLayer object's Recordset object can be done with the following few simple steps:

1. Ensure the existence of a key field. This refers to a field that exists in both the Table object and the Recordset object, and that contains values unique for those fields within each object. The key field in each table can be used to create a "relate" in MapObjects, which is essentially a table join, as described in the "Database Tables" section of Chapter 6.

   Recalling the previous example of the "Schools" Recordset object, the recordset contains a field named SchoolID that contains a number uniquely identifying each school in the Recordset object. Suppose a database table named SchoolData also exists. Further suppose that in addition to having fields containing school population, level, and budget, the SchoolData table has a field named SchoolID, with values identical to those of the Recordset object. In this case, SchoolID can serve as the key field. These conditions are shown in the following illustration.

*An example of a Recordset and Table that can be related.*

"Schools" Recordset

| SchoolName | SchoolID |
|------------|----------|
| Esperanza  | 1        |
| Rose       | 2        |
| Glen Knoll | 3        |

"SchoolData" dBase Table

| SchoolID | Level  | Pop.  |
|----------|--------|-------|
| 1        | High   | 2,000 |
| 2        | Middle | 1,000 |
| 3        | Grade  | 500   |

2. Create an ODBC data source if one does not already exist.

**3.** Set the Database property of the Table object to the ODBC data source containing the table to be represented by the Table object.

**4.** Specify the Name property of the Table object as the name of the external table to be referenced.

**5.** Return a reference to the MapLayer's Recordset.

**6.** Use the AddRelate method of the MapLayer object to establish the relate. AddRelate requires that you specify the key field for the MapLayer's Recordset, the key field for the external table, and the relevant Table object.

In the case of the Recordset of the Schools layer and the table named SchoolsData—both of which contain the key field named SchoolID—the following code would create a relate between the Schools Recordset object and the SchoolsData table. This example assumes that the School-Data table is a dBASE table that can be accessed by an ODBC data source named dBASE Files. The illustration that follows shows the Schools Recordset object after relating the SchoolData table object.

```
Dim aTable As MapObjects.Table
aTable.Database = "dBASE Files" 'the ODBC data source name
aTable.Name = "SchoolData" 'the name of the table
Map1.Layers("Schools").AddRelate("SchoolID", aTable, "SchoolID") ' create the
relate
```

*The Schools Recordset object after relating the SchoolData Table object.*

| SchoolID | SchoolName | Level | Pop. |
|----------|------------|-------|------|
| 1 | Esperanza | High | 2,000 |
| 2 | Rose | Middle | 1,000 |
| 3 | Glen Knoll | Grade | 500 |

# Geometric Objects

*Geometry provides the foundation for spatial analysis.*

*Overlaying polygons is an example of spatial analysis.*

*Spatial analysis requires geometric and locational information.*

*Spatial analysis requires spatial operators.*

Performing queries based on geographic location (also known as spatial analysis) requires the use of geometry. For example, a common analytical method in the world of paper maps is to overlay two maps on a light table to identify where there are overlapping polygons representing areas of interest. These maps might represent data as diverse as land use, soil type, trade areas, emergency response zones, or species conservation districts.

As an application of this type of analysis, consider the need to know whether or not the trade areas for two companies overlap, meaning that the companies are drawing customers from a common area and therefore may be competing with each other. This analysis can be accomplished by overlaying a map of company A's trade areas with a map of company B's trade areas, looking for overlapping polygons, and then analyzing the polygons formed by the overlap to determine their characteristics.

This type of analysis, based on overlapping shapes, requires the use of geometry and geometric shapes to identify and analyze the overlap. The same is true if the analysis involves intersecting shapes (such as a proposed road and impacted parcels), buffered shapes (such as a specified area around a construction site), or contained shapes (such as customer addresses contained within a specified trade area).

Two things are necessary to enable spatial analysis. First, a means must be provided for representing and storing the locational and geometric information (the spatial data) regarding geographically located features such as roads, trade areas, and conservation districts. Second, operators must be provided to perform spatial analysis using the locational and geometric information. Support of these two requirements by MapObjects is discussed in the following sections.

## Representing Geometric Shapes

Geometric shapes can represent the locational and geometric information of real-world objects. For instance, a polygon representing a trade area has geometric information that specifies the actual shape of the area, and locational information that specifies where the shape is located in geographic coordinate space (such as latitude and longitude). MapObjects provides geometric objects as a way to construct and manipulate geometric shapes.

*The Geometric objects of MapObjects.*

MapObjects provides the following seven geometric objects for representation of shapes. These are depicted in the illustration that follows.

- ❏ The *Rectangle* object represents a geometric shape with four edges and four right angles.
- ❏ The *Ellipse* object represents an elliptical geometric shape.
- ❏ The *Polygon* object represents a geometric shape having three or more points.
- ❏ The *Line* object represents a geometric shape having two or more points.
- ❏ The *Point* object represents a geometric shape having one point.
- ❏ The *Points* object is a collection containing all Point objects representing a Line or Polygon object.
- ❏ The *Parts* object is a collection containing all Points collections representing an object that must be represented by multiple shapes. For example, the country of Japan could be represented by a Parts collection. It is a single object (a country) that contains multiple polygonal shapes (the islands), with each polygonal shape possessing a Points collection containing all Point objects for that shape.

*MapObjects
Geometric objects.*

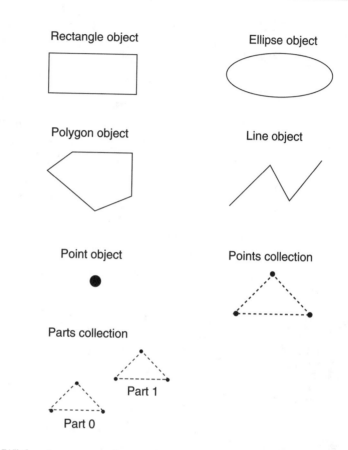

Rectangle object

Ellipse object

Polygon object

Line object

Point object

Points collection

Parts collection

Part 1

Part 0

<u>Geometric objects have
dimensional properties.</u>

With the exception of Points and Parts collections—which only have a Count property to indicate the number of objects in the collection—each geometric object has dimensional properties. These vary depending on the object, but include such properties as Area, Perimeter, Height, Width, Top, Left, Bottom, Right, and Center.

## Analyzing Relationships Among Geometric Shapes

On many occasions the relationships between geometric shapes must be examined. For instance, you may need to determine where a line is intersected by another line, as in the case of determining where a stream crosses a road.

You may also want to determine whether a point is located within a polygon, or what the distance is between two point objects, as in the case of determining the nearest retail store to a particular customer's home.

MapObjects provides methods along with geometric objects to help determine relationships such as these. A number of these methods are common to most of the geometric objects, such as the DistanceTo method for determining the distance between two objects, and the GetCrossings method for identifying the locations at which two objects cross. Other methods are specific to the characteristics of the shape represented by a geometric object, such as the DistanceToSegment method of the Point object for determining the distance from a single Point object to a line.

## Using Shapes, Tables, and Recordsets in Queries

The geometric object methods just discussed have as their express purpose the determination of the relationship between or among spatial features (i.e., spatial analysis). However, it is often necessary to use spatial relationships to query data in a way that returns a subset of features or records.

Recall the example from earlier in this chapter in which a retailer needed to determine which customers live within a specified trade area and also meet certain criteria regarding age and income. As noted in the example, a query of this type involves both a spatial dimension and an aspatial dimension.

This type of analysis brings together the concepts of Recordset objects and geometric objects in order to perform a sophisticated query that can retrieve a subset of the records associated with a MapLayer. In MapObjects, the results of this type of query are a Recordset object with records meeting the criteria specified in the query. MapObjects supplies three methods for performing queries of this type: SearchByDistance, SearchExpression, and SearchShape.

## The SearchByDistance Method

*The SearchByDistance method determines proximity and attribute.*

The *SearchByDistance* method creates a Recordset based on a search for all features within a distance of a specified shape. An expression in the form of an SQL "where" clause can also be supplied to further define the criteria for selection. For example, to locate all contaminated wells in a hydrology layer (named AllWells) within 2,000 feet of a specified point, the following code could be written:

```
Set recset = Map1.Layers("AllWells").SearchByDistance(pnt, 2000, "where condition = 'contaminated'")
```

In the parameter list contained in the parentheses following the `SearchByDistance` method, `pnt` is a variable referencing a MapObjects Point object, `2000` is the specified search distance around the point, and the quoted string is the expression applied to the attributes of the records. Condition is the name of the field indicating whether the well is contaminated. If no expression is supplied, meaning that a spatial search will be the sole criterion for selection, the quoted string should be blank: " ".

## The SearchExpression Method

*The SearchExpression method searches by attribute only.*

The SearchExpression method differs from the SearchByDistance method in that it accepts only an SQL "where" clause expression as the search criterion. No option for spatial searching is provided. To search for contaminated wells in the AllWells layer, regardless of where they are located, the following code could be written:

```
Set recset = Map1.Layers("AllWells").SearchExpression("where condition = 'contaminated'")
```

## The SearchShape Method

The SearchShape method creates a Recordset containing records meeting the criteria specified by a spatial operator supplied by MapObjects. Additionally, an SQL expression can be supplied. The SearchShape method differs from

The SearchShape method allows complex spatial and attribute searches.

the SearchByDistance method in that SearchByDistance only supports a proximity search; that is, features within a given distance of a specified location. SearchShape, on the other hand, supports complex spatial operations. For instance, to create a Recordset object containing all land parcels intersecting a proposed road, the following code could be written:

```
Set recset = Map1.Layers("TheParcelLayer").SearchShape(aRoadLineShape, moLine-
Cross, "")
```

Here, aRoadLineShape is a MapObjects Line object representing the road, and moLineCross is a MapObjects spatial operator constant. In this case, moLineCross specifies that a spatial search be performed for features from TheParcelLayer that intersect (cross) the Line object representing the road. Note that no SQL expression is supplied.

## MapObjects Spatial Operators

MapObjects supports a variety of spatial operators.

MapObjects supplies the following 15 spatial operators, which are identified by their constants.

❑ *moExtentOverlap* returns features whose extents overlap the extent of the search feature.

❑ *moCommonPoint* returns features that share at least one identical common point with the search feature.

❑ *moLineCross* returns features that intersect the search feature.

❑ *moCommonLine* returned features must share at least one identical common line segment with the search feature.

❑ *moCommonPointOrLineCross* returns features that share a common point with the search feature or that intersect it.

❑ *moEdgeTouchOrAreaIntersect* returns features that touch the search feature, are wholly or partially within the search feature, or wholly or partially contain the search feature(s).

☐ *moAreaIntersect* returns features that are wholly or partially contained within it, but not adjacent to it, if the search feature is a polygon feature. Otherwise, the features themselves must be polygon features, and the method returns features that wholly or partially contain the search feature.

☐ *moAreaIntersectNoEdgeTouch* is the same as moAreaIntersect, but the boundaries of the search feature and the feature may not intersect or touch.

☐ *moContainedBy* returns features that wholly contain the search feature. If the feature is a polygon feature, the search feature must be wholly inside it, inclusive of the feature's boundary. If the feature is a line feature, the search feature must lie along the feature's path. If the feature is a point feature, the search feature must be on one of its vertexes.

☐ *moContaining* returns features that are wholly contained within the search feature.

☐ *moContainedByNoEdgeTouch* returns features that wholly contain the search feature, not inclusive of the search feature's boundary. The feature must be a polygon feature, the search feature must be wholly inside it, and their boundaries may not intersect or touch.

☐ *moContainingNoEdgeTouch* returns features that are wholly within the search feature, not inclusive of the search feature's boundary. The search feature must be a polygon feature, the feature must be wholly inside it, and their boundaries may not intersect or touch.

☐ *moPointInPolygon* returns polygon features that contain the first coordinate of the search feature.

☐ *moCentroidInPolygon* returns polygon features whose centroids are contained by the shape.

❏ *moIdentical* returns features that are identical to the search feature. Considers feature type and coordinate description. Typically used to find duplicate data.

A Visual Basic application named Spatial.vbp is included in the Visual Basic sample applications on the CD-ROM accompanying this book. This application provides examples of the use of most of these 15 spatial operators.

## The Describe Shapefile Application: Recordsets, Shapes, and Queries

Among the Visual Basic sample applications on the CD-ROM included with this book, you will find a Visual Basic project named DescribeShapefile.vbp. If you indicated when installing MapObjects from the CD-ROM that you wanted the samples installed, the Visual Basic applications are available in the *samples\vb* directory located in the directory in which MapObjects was installed on your machine. The DescribeShapeFile project allows the user to browse information, such as field names and definitions, or an ESRI shapefile. This section walks through the DescribeShapefile project to demonstrate the use of recordsets, geometric shapes, and queries in an actual application.

Although you can simply read this section to gain an understanding of the application of recordsets and geometric objects, you are encouraged to open the project and follow along. As with the examples in the previous chapters, you can run the code as it stands, and because the source code is provided you can also get hands-on experience by changing values and running the code to see the effect of your changes.

### The DescribeShapefile Application's User Interface

The primary user interface of the DescribeShapefile application is contained in three forms: Form1 (named Describe.frm), Form2 (named View.frm), and Form3 (named Identify.frm). Form1, which is defined in the project as the form to be presented at startup, consists of the following elements. The illustration that follows shows the Form1 interface at run time.

❏ A *Describe Shapefile...* button (named Command1) that when pressed opens a dialog box allowing the user to select a shapefile, then retrieves descriptive information about the shapefile, and then places the information in the other controls on Form1 for presentation.

❏ A TextBox named *Text1* placed just below the Shapefile label (named Label2), which displays the name of the selected shapefile.

❏ A TextBox named *Text2* placed just below the Shape Type label (named Label3), which displays a description of the type of features contained in the shapefile (i.e., point, line, or polygon).

❏ A TextBox named *Text3* placed just below the Number of Records label (named Label4), which displays the number of records in the shapefile's recordset.

❏ A ListView named *ListView1* placed just below the Fields label (named Label1), which displays the name and type of each field in the shapefile's recordset.

❏ A *View...* button (named Command2) that when pressed will display another form (Form2) in which to display a view of the shapefile's geography in a Map control.

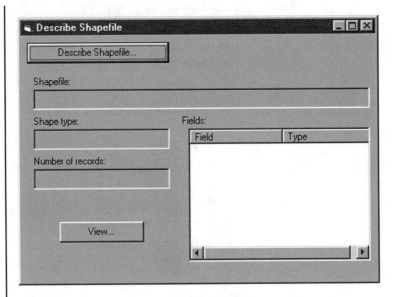

*The user interface of Form1 of the DescribeShapefile project as seen at run time prior to selecting a shapefile.*

In addition to these controls, which are visible at run time, is the following control on Form1, which is only visible at design time. The illustration that follows shows Form1 at design time.

❑ A *CommonDialog control* (named CommonDialog1) that is used to generate the dialog box opened by the Shapefile... (Command1) button.

The user interface of Form2 (View.frm), which is displayed by clicking on the Form1 (Describe.frm) Command2 (View...) button, consists of the following elements. The illustrations that follow show the user interface of Form2 at run time, and Form2 and its controls at design time.

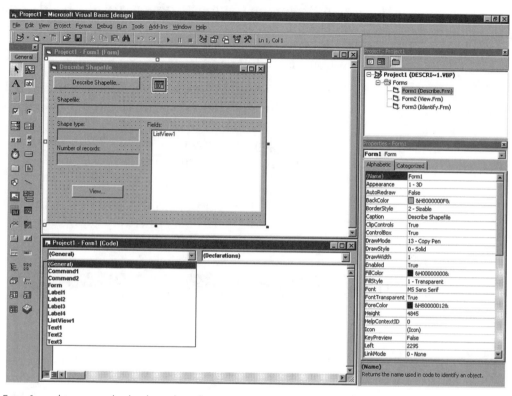

*Form1 and its controls displayed at design time.*

❏ A Map control named *Map1* on which the view of the shapefile's geography is displayed.

❏ A Toolbar control named *Toolbar1* that allows the user to specify what actions should occur when clicking on the map with the mouse. The toolbar contains the following buttons:

• A *Zoom* button (button 1 in the toolbar), displayed with a magnifying glass icon. This button enables the user to zoom in to an area on the map defined by a clicked-and-dragged box.

- A *Pan* button (button 2), displayed with a hand icon. This button allows the user to drag the map in any direction to display areas currently outside the bounds of the map.

- An *Identify* button (button 3), displayed with a circled I icon. After the user selects this button, clicking on the map will display another form (Form3), which displays data contained in the attribute record of the map feature clicked on.

- A *Full Extent* button (button 4), displayed with a globe icon. This button sets the map view to the full extent of the layer. This effectively zooms the view as far out as the map will allow.

*The user interface of Form2 of the DescribeShapefile project as seen at run time after selecting a shapefile.*

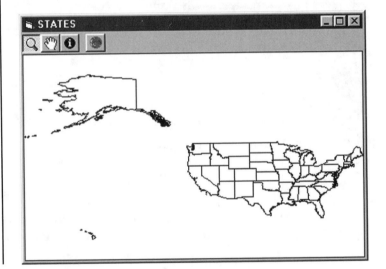

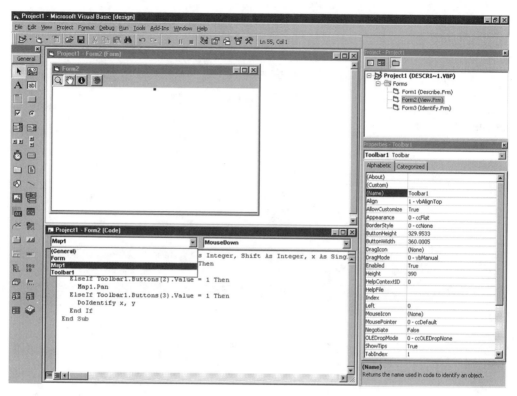

*Form2 and its controls displayed at design time.*

The user interface of Form3 (Identify.frm), which is displayed by clicking on the Form2 (View.frm) Toolbar1 Identify button, contains a single control: ListView1. The illustrations that follow show the user interface of Form3 at run time, and Form3 displayed at design time.

❒ The ListView control named *ListView1* displays the values of the attribute record of the last map feature clicked on.

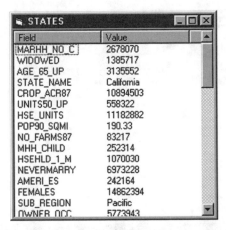

*The user interface of Form3 of the DescribeShapefile project as seen at run time after clicking on a map feature.*

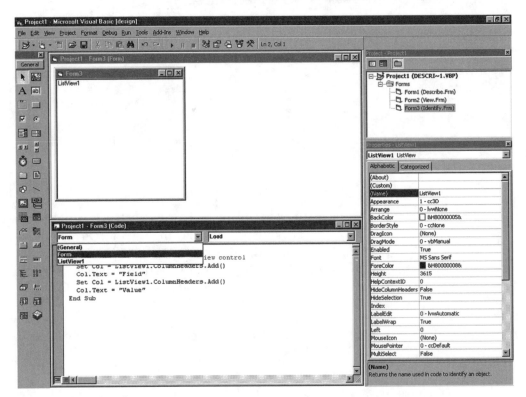

*Form3 displayed at design time.*

## Form1 General Declarations: Making the Shapefile Accessible

The General Declarations section of a form provides a place where, among other things, variables can be declared in order to make them accessible in multiple procedures. The code for the General Declarations section of Form1 includes the following single line of code:

```
Dim g_layer As MapLayer
```

The variable `g_layer` is declared in this line of code as a MapLayer object variable. Declaring the variable in the General Declarations section of the form makes it available to all procedures in Form1. When this variable is later set to reference a shapefile, the shapefile will then be accessible to every procedure in Form1.

## The Form1 Form Load Procedure: Preparing the User Interface

The Load event of Form1 occurs when the Describe-Shapefile application is started. This is because the DescribeShapefile project is set to automatically open Form1 when executed. Opening the form at execution causes the form's Load event to occur. When this event occurs in the DescribeShapefile application, the following program code, contained in Form1's Form Load procedure, is executed:

```
Private Sub Form_Load()
 ' set up the list of fields
 Set Col = ListView1.ColumnHeaders.Add()
 Col.Text = "Field"
 Set Col = ListView1.ColumnHeaders.Add()
 Col.Text = "Type"
End Sub
```

This procedure modifies `ListView1` (which will eventually display the name and type of each field in the shapefile's recordset) by adding two column headers named

`Field` and `Type`. The following illustration shows the Form1 ListView object at design time, and at run time after the Form Load procedure.

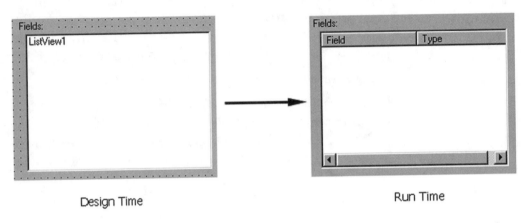

Design Time                                        Run Time

*The ListView object of Form1 (Describe.frm) as it appears at design time, and at run time after executing the Form1 Form Load procedure.*

# The Form1 Command1 Button:
# Selecting the Shapefile and Displaying Its Description

The Command1 button (labeled Describe Shapefile... in the user interface) will when pressed display a dialog box that allows you to select a shapefile to describe. When a selection is made in the dialog box, the dialog box is closed and Form1 displays descriptive information regarding the shapefile, as shown in the following illustration.

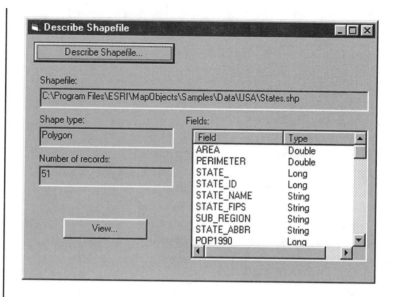

*Form 1 (Describe.frm) as it appears at run time after clicking the Command1 (Describe Shapefile...) button and selecting a shapefile.*

The program code that accomplishes this is contained in the Command1 button's Click procedure. When the Command1 button is pressed, the button's Click event occurs. When the Click event occurs, the Command1 Click procedure runs and executes the following program code:

```
Private Sub Command1_Click()
 CommonDialog1.Filter = "ESRI Shapefiles (*.shp)|*.shp"
 CommonDialog1.ShowOpen
 If Len(CommonDialog1.filename) = 0 Then Exit Sub

 Unload Form2

 Dim dc As New DataConnection
 dc.Database = CurDir
 If Not dc.Connect Then Exit Sub

 Text1.Text = CommonDialog1.filename

 Dim name As String
```

```
name = Left(CommonDialog1.FileTitle, Len(CommonDialog1.FileTitle)    - 4)
Dim gs As GeoDataset
Set gs = dc.FindGeoDataset(name)
If gs Is Nothing Then Exit Sub


Set g_layer = New MapLayer
Set g_layer.GeoDataset = gs


If g_layer.shapeType = moPolygon Then
 Text2.Text = "Polygon"
 g_layer.Symbol.Color = moLightYellow
ElseIf g_layer.shapeType = moLine Then
 Text2.Text = "Line"
 g_layer.Symbol.Color = moDarkGreen
Else
 Text2.Text = "Point"
 g_layer.Symbol.Color = moRed
End If


Dim recs As MapObjects.Recordset
Set recs = g_layer.Records
Text3.Text = recs.Count


Dim desc As TableDesc
Set desc = recs.TableDesc


ListView1.ListItems.Clear
For i = 0 To desc.FieldCount - 1
 Dim s As String
 Select Case desc.FieldType(i)
  Case moBoolean
   s = "Boolean"
  Case moLong
   s = "Long"
  Case moDate
   s = "Date"
  Case moDouble
   s = "Double"
  Case moString
```

```
  s = "String"
 Case moNone
  s = "Invalid"
 End Select

 Set newItem = ListView1.ListItems.Add
 newItem.Text = desc.fieldName(i)
 newItem.SubItems(1) = s
 Next i

End Sub
```

## Selecting the Shapefile

The following is the first block of code in this procedure. It provides the user with a shapefile selection dialog box by using the CommonDialog control. The CommonDialog control provides a standard set of dialog boxes for operations such as opening and saving files, setting print options, and selecting colors and fonts.

```
CommonDialog1.Filter = "ESRI Shapefiles (*.shp)|*.shp"
CommonDialog1.ShowOpen
If Len(CommonDialog1.filename) = 0 Then Exit Sub

Unload Form2
```

The first line of this block of code sets the file filter of the dialog box that will be displayed to show only ESRI shapefiles (files ending with the extension *.shp), as shown in the following illustration. The second line displays the dialog box for the user by invoking the ShowOpen method of the CommonDialog control.

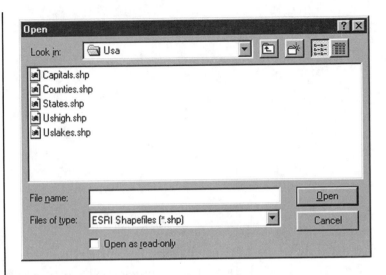

*The shapefile selection dialog box displayed at run time.*

After the user has made a selection from the dialog box, control is returned to this procedure at the line containing the `If` statement in the previous block of code shown here. The `If` statement evaluates the Filename property of the CommonDialog control, which contains the path and filename of the selected file. The `If` statement will exit this procedure when a zero-length value for the Filename property indicates that no file was selected in the dialog box.

The last line of this block of code unloads `Form2` (View.frm), removing the form from display onscreen. This is done because selection of a new shapefile will invalidate any map view currently presented in `Form2`.

## Creating the DataConnection

The next block of code creates a DataConnection object named `dc` that establishes the data source as the directory in which the selected shapefile is contained.

```
Dim dc As New DataConnection
dc.Database = CurDir
If Not dc.Connect Then Exit Sub
```

The first line creates a new `DataConnection` object. The second line associates it (via the `Database` property) to the directory from which the shapefile was selected. The Visual Basic `CurDir` function is used to specify this directory. `CurDir` returns a string indicating the current directory.

When a shapefile was selected using the dialog box, the current directory was set to the directory containing the shapefile. The `If` statement in the third line will exit the procedure if the establishment of the `DataConnection` object was unsuccessful. The following is the next line in the procedure:

```
Text1.Text = CommonDialog1.filename
```

This line begins the display of descriptive information by setting the `Text` property of the `Text1` TextBox control to display the full path and filename of the selected shapefile in the TextBox. The following illustration shows this display at run time.

*Displaying a shapefile's full path and filename at run time in the Text1 Textbox of Form1.*

## Creating the GeoDataset

The next block of code within the Command1 Click procedure creates a GeoDataset object named gs to reference the selected shapefile within the data source now identified by the DataConnection object.

```
Dim name As String
 name = Left(CommonDialog1.FileTitle, Len(CommonDialog1.FileTitle) - 4)
 Dim gs As GeoDataset
 Set gs = dc.FindGeoDataset(name)
 If gs Is Nothing Then Exit Sub
```

The first two lines set the string variable name to the filename (without the path) of the selected shapefile. Notice the use of the Visual Basic Left function in the second line to extract a specified number of characters from the filename. The first parameter in the parentheses after the Left function, CommonDialog1. FileTitle, is the shapefile's filename identified by the FileTitle property of the CommonDialog control. The FileTitle property contains the filename of the selected file without the path.

The second parameter in the parentheses after the Left function, Len(CommonDialog1.FileTitle) – 4, is the number of characters to extract. This number is identified as four (4) less than the total length of the filename (returned by the Visual Basic Len function), in order to remove the file extension. The result of this line of code is that the variable name now contains the shapefile name without a pathname and .shp extension.

The last three lines in this block of code declare a GeoDataset object variable named gs, set the variable to reference the GeoDataset identified by name, and check that the reference was successful. If the If statement in the last line determines that the reference was unsuccessful (i.e., gs Is Nothing), the procedure is exited.

### Creating the MapLayer

The next two lines of code in the Command1 button Click procedure create a new MapLayer object and assign it to the variable g_layer, as follows:

```
Set g_layer = New MapLayer
Set g_layer.GeoDataset = gs
```

The first line creates and assigns the `MapLayer` object. The second line associates the `GeoDataset` representing the selected shapefile to the new `MapLayer` object. Recall that the variable `g_layer` was declared in the General Declarations section of the form in order to make it accessible to every procedure in the form. Therefore, assigning the selected shapefile as a `MapLayer` object referenced by `g_layer` makes the selected shapefile accessible to all procedures in the form. Now that the shapefile is accessible and assigned to a MapLayer, you can begin determining and displaying its characteristics.

## Determining and Displaying the ShapeType

In the next block of code in the Command1 Click procedure, the If statement determines the type of geometric shape (e.g., point, line, or polygon) associated with the MapLayer object:

```
If g_layer.shapeType = moPolygon Then
  Text2.Text = "Polygon"
  g_layer.Symbol.Color = moLightYellow
ElseIf g_layer.shapeType = moLine Then
  Text2.Text = "Line"
  g_layer.Symbol.Color = moDarkGreen
Else
  Text2.Text = "Point"
  g_layer.Symbol.Color = moRed
End If
```

This code references the ShapeType property of the MapLayer object to determine the type of geometric objects contained in the MapLayer object's GeoDataset. The first portion of the `If` statement determines, using the MapObjects constant `moPolygon`, whether or not the ShapeType property indicates that the geometry of the MapLayer's GeoDataset consists of polygons.

```
If g_layer.shapeType = moPolygon Then
  Text2.Text = "Polygon"
  g_layer.Symbol.Color = moLightYellow
```

If this is true, the Text property of the `Text2` TextBox is set to display the word *Polygon* in the TextBox. The last line sets the color of the layer to light yellow using the appropriate MapObjects color constant.

The second portion of the `If` statement follows. This portion executes if the first `If` statement evaluated as false (i.e., the value of the ShapeType property did not indicate polygonal geometry).

```
ElseIf g_layer.shapeType = moLine Then
  Text2.Text = "Line"
  g_layer.Symbol.Color = moDarkGreen
```

The `ElseIf` determines, using the MapObjects constant `moLine`, whether or not the ShapeType property indicates that the geometry of the MapLayer's GeoDataset consists of lines. If this is true, the `Text` property of the `Text2` TextBox is set to display the word `Line` in the TextBox. The last line sets the color of the layer to dark green using a MapObjects color constant.

The last portion of the `If` statement follows. It executes if both the `If` statement and the `ElseIf` statement evaluated to false (i.e., the value of the ShapeType property did not indicate polygonal or line geometry).

```
Else
  Text2.Text = "Point"
  g_layer.Symbol.Color = moRed
End If
```

The only remaining option in this case is that the Shape-Type property indicates point geometry. Therefore, the Text property of the `Text2` TextBox is set to display the word *Point* in the TextBox. The last line sets the color of the layer to red using a MapObjects color constant.

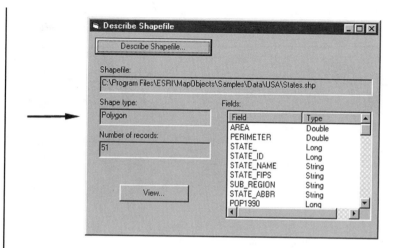

*Displaying a shapefile's shape type at run time in the Text2 Textbox of Form1.*

## Accessing the Recordset Object and Displaying the Number of Records

The next block of code follows. It first declares a variable named recs as a MapObjects Recordset object.

```
Dim recs As MapObjects.Recordset
Set recs = g_layer.Records
Text3.Text = recs.Count
```

The second line sets this variable to reference the Recordset object of the MapLayer (g_layer), which represents the selected shapefile. This is done by referencing the Records property of the MapLayer object in order to return the associated Recordset object. The final line in this block sets the Text property of the Text3 TextBox to display the number of records in the Recordset object, as shown in the following illustration. The number of records is returned by referencing the Count property of the Recordset object.

*Displaying the number of records at run time in the Text3 Textbox of Form1.*

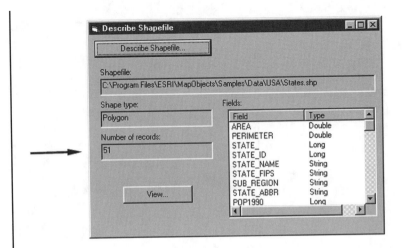

## Accessing the TableDesc and Field Objects

The remaining code in the Command1 Click procedure accesses the TableDesc and Field objects and displays the properties to the user via the form. The first two lines follow.

```
Dim desc As TableDesc
Set desc = recs.TableDesc
```

These two lines declare a variable named `desc` as a `TableDesc` object and set the variable to reference the `TableDesc` object of the Recordset `recs` (the recordset of the selected shapefile). Recall that a `TableDesc` object contains a description of the individual Field objects of a Recordset object. The next several lines write descriptive information, retrieved from the TableDesc and Field objects, to Form1 for display to the user.

```
ListView1.ListItems.Clear
 For i = 0 To desc.FieldCount - 1
  Dim s As String
  Select Case desc.FieldType(i)
   Case moBoolean
    s = "Boolean"
```

```
      Case moLong
        s = "Long"
      Case moDate
        s = "Date"
      Case moDouble
        s = "Double"
      Case moString
        s = "String"
      Case moNone
        s = "Invalid"
    End Select

    Set newItem = ListView1.ListItems.Add
    newItem.Text = desc.fieldName(i)
    newItem.SubItems(1) = s
  Next i

End Sub
```

The first line clears any existing information out of the ListView object's display on Form1. The second line is a For...Next statement beginning a block of code that will execute once for each field in the TableDesc object named `desc`.

The third line declares a string variable named `s` that will hold a description of the type of data stored in the field currently being processed by the For...Next statement. The fourth line begins a `Select Case` block that will evaluate the current field to determine its data type, as follows:

```
Select Case desc.FieldType(i)
    Case moBoolean
      s = "Boolean"
    Case moLong
      s = "Long"
    Case moDate
      s = "Date"
    Case moDouble
```

```
      s = "Double"
   Case moString
      s = "String"
   Case moNone
      s = "Invalid"
   End Select
```

The `Select Case` block executes one of several statements, depending on the value of `desc.FieldType(i)`. In this expression, the `FieldType` property of the TableDesc object returns the data type of the current field.

Recall that the For...Next block within which this `Select Case` block is contained executes once for every Field object in the TableDesc object associated with the selected shapefile. Because `i` is incremented each time the For statement executes, `i` represents an index to the current Field object. This allows you to use `i` as a way of evaluating every field once with this `Select Case` statement.

The `Case` statements evaluate the `FieldType` property of the current field by comparing it to MapObjects FieldType constants (e.g., `moLong` and `moString`). The appropriate text string is assigned to the string variable `s` according to the results of the evaluation.

The following are the final lines of the Command1 Click procedure. They complete the For...Next block.

```
Set newItem = ListView1.ListItems.Add
   newItem.Text = desc.fieldName(i)
   newItem.SubItems(1) = s
 Next i


End Sub
```

A variable named `newItem` is set to reference a ListItem added to the ListView's ListItem collection using the `Add` method. The `Text` property of the new ListItem is set to display the name of the current field by referencing the FieldName property. This text will display under the Field column header defined in the Form1 Form Load procedure. Additional text is displayed by referencing the variable s, which contains the string describing the data type contained in this field. This text will display under the Type column header defined in the Form1 Form Load procedure.

The Next statement continues the For...Next block's processing until all fields have been processed. Then the Command1 Click procedure exits. The result is a display in Form1 of descriptive information relating to the fields of the selected shapefile, as shown in the following illustration.

*Displaying a shapefile's field descriptions at run time in the ListView1 ListView of Form1.*

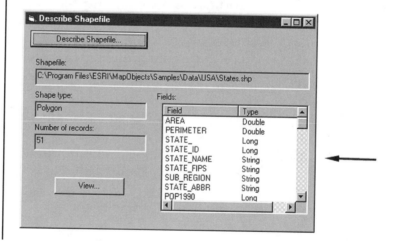

# The Form2 Command2 Button:
# Viewing the Geography of the Shapefile

The Command2 button (labeled View... in the user interface) will when pressed display another form, Form2 (View.frm), to display a map of the geography of the selected shapefile. This is shown in the following illustration.

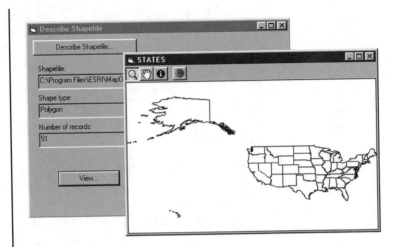

*The application as it appears at run time with Form2 (View.frm) displayed. Form2 was displayed here after selecting a shapefile in Form1 (Describe.frm) and clicking the Form1 Command2 (View...) button.*

The program code that displays Form2 is found in the Form1 Command2 button's Click procedure. When the Command2 button is pressed, the button's Click event occurs. When the Click event occurs, the Command2 Click procedure runs and executes the following program code:

```
Private Sub Command2_Click()
  If (Not g_layer Is Nothing) And (Not Form2.Visible) Then
    Form2.Map1.Layers.Clear
    Form2.Map1.Layers.Add g_layer
    Form2.Map1.Extent = Form2.Map1.FullExtent
    Form2.Caption = g_layer.name
    Form2.Show
  End If
End Sub
```

The code in this procedure consists of an If...Endif statement containing the code required to set up and display Form2. The initial If statement determines whether or not the variable g_layer references a valid MapLayer *and* Form2 is visible. If a shapefile is referenced and Form2 is not currently displayed, the block of code within the If...Endif statement executes.

The first line of code within the If...Endif statement follows. It invokes the Clear method of the Layers collection of the Map1 control contained in Form2.

```
Form2.Map1.Layers.Clear
```

Notice the full qualification of the location of the Map control named Map1. It is located on Form2, but this code is in Form1. Therefore, the map is explicitly referenced as Form2.Map1. The Clear method removes all members of the Layers collection, which essentially removes all layers from Map1.

The second and third lines of code follow. These lines invoke the Add method of the Layers collection to add the selected shapefile to Map1's Layers collection, and set the Extent property of the Map control. The Extent property of the Map control is set to the Rectangle object representing the full extent of all of the Map control's associated MapLayer objects (which is currently only g_layer).

```
Form2.Map1.Layers.Add g_layer
Form2.Map1.Extent = Form2.Map1.FullExtent
```

The fourth and final lines of code in this block complete the setup and display of Form2, as follows:

```
Form2.Caption = g_layer.name
Form2.Show
```

First, the Caption property (which controls the text displaying in a form's title bar) is set to the name of the shapefile. Then the Show method of Form2 is invoked to display the form on the screen. There is no program code in the Form2 Form Load procedure. Therefore, the form loads without any additional processing, and a map of the selected shapefile is displayed in the Map1 control of Form2.

# The Form2 Map1 MouseDown Event: Handling a Click on the Map

When the user performs a mouse click on Form2's Map control, named Map1, the MouseDown event occurs and the following code from the Map1 MouseDown procedure executes:

```
Private Sub Map1_MouseDown(Button As Integer, Shift As Integer, x As Single, y As Single)
 If Toolbar1.Buttons(1).Value = 1 Then
  DoZoom
 ElseIf Toolbar1.Buttons(2).Value = 1 Then
  Map1.Pan
 ElseIf Toolbar1.Buttons(3).Value = 1 Then
  DoIdentify x, y
 End If
End Sub
```

Notice first that the MouseDown event is defined by MapObjects to receive four parameters: an integer value named `Button`, an integer value `Shift`, and two single precision numbers named `x` and `y`. (See Chapter 6 for a list of all events recognized by MapObjects and explanations of their available arguments.)

The If statements in the Map1 MouseDown procedure determine what button is currently selected in the toolbar (`Toolbar1` of Form2). For example, if button 1 (the Zoom In button) is selected, the DoZoom procedure is executed.

## *The Toolbar1 Zoom Button: Using the Map Extent to Zoom In*

If button 1 (the Zoom button) of Toolbar1 is currently selected when the MouseDown event occurs on Map1, the following code from the Map1 MouseDown procedure executes:

```
If Toolbar1.Buttons(1).Value = 1 Then
  DoZoom
```

This will execute the `DoZoom` procedure. The `DoZoom` procedure is a general procedure located in the General section of Form2. The `DoZoom` procedure contains the following code:

```
Sub DoZoom()
  ' get a rectangle from the user
  Set r = Map1.TrackRectangle
  ' zoom to the rectangle if its valid
  If Not r Is Nothing Then Map1.Extent = r
End Sub
```

The code first sets a variable named `r` to reference a Rectangle object returned from the `TrackRectangle` method of the Map control. Recall from previous examples that the `TrackRectangle` method allows the user to drag and draw a rectangular shape on the Map control.

The code then determines whether or not a valid rectangle was returned (`If Not r Is Nothing`). If a rectangle was returned, the `Extent` property of the Map control is set to the Rectangle object referenced by `r`. The effect of this code is to zoom in on an area defined by the user.

## The Toolbar1 Pan Button: Using the Map.Pan Method

If button 2 (the Pan button) of Toolbar1 is currently selected when the MouseDown event occurs on Map1, the following code from the previously listed Map1 Mouse-Down procedure executes:

```
ElseIf Toolbar1.Buttons(2).Value = 1 Then
  Map1.Pan
```

This will execute the `Pan` method of the Map control. The `Pan` method allows the user to move the map by clicking on the map and then moving the mouse while holding the mouse key down. As described in previous examples, this will pan the map (move it in any direction) according to the user's mouse movements.

## The Toolbar 1 Identify Button: Performing Spatial Queries

If button 3 (the Identify button) of Toolbar1 is currently selected when the MouseDown event occurs on Map1, a query is performed to determine the map feature on which the user clicked, the feature's attributes are retrieved, and Form3 (Identify.frm) is displayed to present the attributes to the user, as shown in the following illustration.

*The application as it appears at run time with Form3 (Identify.frm) displayed.*

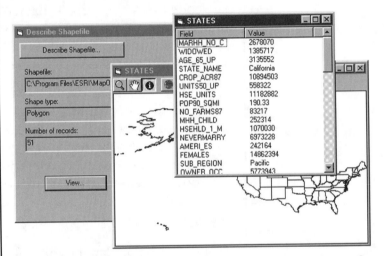

Form3 was displayed in the previous illustration after selecting the Identify button on the toolbar of Form2 (View.frm), then clicking on the state of California in the U.S. map displayed on Form2. The Map control detects and responds to the selection of button 3 in conjunction with a mouse click on the map in the following code from the Map1 MouseDown procedure:

```
ElseIf Toolbar1.Buttons(3).Value = 1 Then
  DoIdentify x, y
```

If button 3 is selected, this code executes the DoIdentify procedure, which is a general procedure located in the General section of Form2. Notice that the call to execute DoIdentify passes two parameters to the procedure: x, which is the X coordinate of the location of the mouse click on Map1, and y, which is the Y coordinate of the mouse click on Map1. Recall that these values were passed by the Map control to the MouseDown procedure as output from the MouseDown event. This code is passing these values on to the DoIdentify procedure. The DoIdentify procedure contains the following code:

```
Sub DoIdentify(x As Single, y As Single)
 ' get the layer
 Set l = Map1.Layers(0)
 ' transform the point to map coordinates
 Set p = Map1.ToMapPoint(x, y)


 ' perform the search
 If l.shapeType = moPolygon Then
  Set recs = l.SearchShape(p, moPointInPolygon, "")
 Else
  Set recs = l.SearchByDistance(p, Map1.ToMapDistance(100), "")
 End If


 ' if the search returned something, display the fields
 ' and values
 If Not recs.EOF Then
  ' show the identify window
  Form3.Show
  Form3.Caption = Form2.Caption
  Form3.ListView1.ListItems.Clear


  For Each fld In recs.Fields ' iterate over the fields
   Set newItem = Form3.ListView1.ListItems.Add
   newItem.Text = fld.name
   newItem.SubItems(1) = fld.ValueAsString ' get the value
  Next fld
 End If
End Sub
```

## Preparing References to the Map and the Click Location

The first block of code in this procedure follows. This block sets a variable named l to reference the MapLayer representing the selected shapefile. Because the selected shapefile is the only layer in Map1, it can be referenced by the index 0.

```
' get the layer
Set l = Map1.Layers(0)
' transform the point to map coordinates
Set p = Map1.ToMapPoint(x, y)
```

The last line of this block of code sets a variable p to reference a Point object created using the ToMapPoint method. Recall from previous examples that the ToMapPoint method uses coordinates in control units to create a Point object in map units (e.g., feet). The values for x and y were passed to this procedure from the MouseDown event; therefore, the result is that p represents the location of the user's mouse click on the map.

## Performing Spatial Queries

The next block of code in the DoIdentify procedure follows. An If statement beginning this code checks the ShapeType property of the selected shapefile.

```
' perform the search
If l.shapeType = moPolygon Then
  Set recs = l.SearchShape(p, moPointInPolygon, "")
Else
  Set recs = l.SearchByDistance(p, Map1.ToMapDistance(100), "")
End If
```

If the ShapeType property indicates polygonal geometry (moPolygon), the next line is executed to perform a spatial query appropriate to polygon geometry. This spatial query uses the SearchShape method.

## Querying with SearchShape

Recall from the discussion earlier in this chapter that the SearchShape method returns a Recordset object containing records meeting the criteria specified by a spatial operator selected from those supplied by MapObjects. Additionally, an SQL expression can be supplied. In the following line of code, p is the variable declared earlier in this procedure to reference the location of the mouse click on Map1.

```
Set recs = l.SearchShape(p, moPointInPolygon, "")
```

moLineCross is a MapObjects spatial operator constant that returns the polygon feature containing the Point object referenced by p. Note that no SQL expression is supplied. Therefore, the third and final parameter within the parentheses is a zero-length string ("" ). The result of a successful execution of this line of code is the return of the polygon on which the user clicked.

## Querying with SearchByDistance

If the ShapeType property indicated that the geometry of the MapLayer object referenced by l is not polygonal, the following code executes:

```
Else
    Set recs = l.SearchByDistance(p, Map1.ToMapDistance(100), "")
```

This performs a spatial query appropriate to point or line geometry by using the SearchByDistance method. In this case, you want to find a feature near the location clicked on the map by the user because it is assumed that the user was pointing at a feature.

In this line of code, p, as before, is the variable declared earlier in this procedure to reference the location of the mouse click on Map1. The second parameter enclosed in parentheses is a value indicating the distance from the point to be used in determining which feature to select.

The `ToMapDistance` method is used for this purpose. Similar to the ToMapPoint method used earlier in this procedure, `ToMapDistance` converts a linear measurement in control units to a distance in map units. In this case, the linear distance is 100 control units.

Note that here, as in the previous use of SearchShape, no SQL expression is supplied. Therefore, the third and final parameter within the parentheses is a zero-length string (`""`). The result of a successful execution of this line of code is the return of the first feature within 100 control units of the location the user clicked on the map.

## Preparing Form3 (Identify.frm) to Display the Query Results

The next block of code in the DoIdentify procedure follows. The block begins with an If statement to verify that records were returned from the previous spatial query. If records were returned, you are positioned at a valid record rather than EOF (end of file), the If statement evaluates as true, and processing of the code within the If...Endif statement begins.

```
' if the search returned something, display the fields
 ' and values
 If Not recs.EOF Then
  ' show the identify window
  Form3.Show
  Form3.Caption = Form2.Caption
  Form3.ListView1.ListItems.Clear
```

The three lines of code following the `If` statement reference `Form3` to display it onscreen using the form's `Show` method; set the `Caption` property to display the same text as the caption (title bar) of `Form2`, which is the name of the selected shapefile; and clear all previously displayed text from `Form3`'s ListView object by invoking the `Clear` method.

Note that `Form3` contains only a small bit of code, located in the Form Load procedure, which creates column headers in `Form3`'s ListView object with the titles Field and Value. These columns will be used to display the results of the query. The code in `Form3`'s Form Load procedure is as follows:

```
Private Sub Form_Load()
 ' set up the columns of the listview control
 Set Col = ListView1.ColumnHeaders.Add()
 Col.Text = "Field"
 Set Col = ListView1.ColumnHeaders.Add()
 Col.Text = "Value"
End Sub
```

## Displaying the Query Results

The final block of code in the DoIdentify procedure displays the results of the query in Form3 (Identify.frm) by executing the following code:

```
For Each fld In recs.Fields ' iterate over the fields
   Set newItem = Form3.ListView1.ListItems.Add
   newItem.Text = fld.name
   newItem.SubItems(1) = fld.ValueAsString ' get the value
Next fld
```

This For Each…Next statement repeats the execution of the block of code it contains once for each member of the `Fields` collection associated with the Recordset object (`recs`) of the selected shapefile.

The first line of the block of code within the `For` statement follows. This line sets the variable named `newItem` to reference a new item added to the ListView of `Form3`.

```
Set newItem = Form3.ListView1.ListItems.Add
```

The next line of the block follows. This line sets the Text property of the newItem to the Name property of the Field currently being processed.

```
newItem.Text = fld.name
```

The last line of the block follows. This line sets a SubItem of the ListView to display the value of the Field currently being processed.

```
newItem.SubItems(1) = fld.ValueAsString ' get the value
```

Note the use of the `ValueAsString` method of the Field object. This method returns the value of the field as a string data type.

The Next statement returns control to the For Each statement, which moves to the next record and repeats the previous processing for the new current record. When all records have been processed, the For Each...Next statement is exited. The result of this processing is a form (Form3) displayed onscreen and showing field names and their values for the selected shapefile. See the previous illustration for an example of how Form3 is presented at run time. When this is complete, the end of the DoIdentify procedure is reached and the procedure is exited.

# The Form1 and Form2 Unload Procedures:
# Cleaning Up When Closing Forms

Form1 contains code for the Unload event procedure. The Unload event occurs just prior to a form being removed from the screen. It provides an opportunity to perform cleanup, data validation, or any other activity appropriate to the removal of a form. In the Describe-Shapefile project, the Form1 Unload event procedure contains the following code:

```
Private Sub Form_Unload(Cancel As Integer)
 Unload Form2
End Sub
```

This code removes `Form2` (View.frm) from the screen if it is currently being displayed. `Form2` contains the following code in its Unload event procedure:

```
Private Sub Form_Unload(Cancel As Integer)
 ' get rid of the identify form
 Unload Form3
End Sub
```

This code removes `Form3` (Identify.frm) from the screen if it is currently being displayed. The effect of both of these procedures is to recognize the dependencies between the forms. Because `Form3` is used for displaying feature attributes—which requires the map contained on Form2 (View.frm) for choosing a feature—`Form3` is closed whenever Form2 is closed by the user. Because Form1 (Describe.frm) is the application's start-up form, closing Form1 is equivalent to exiting the application. Therefore, whenever Form1 is closed, Form2 is also closed, which in turn closes `Form3`, resulting in all forms being closed when the application is exited.

# 10

# Looking Ahead at Software Components

**Software components will converge with other technologies.**

It is rarely a single technological innovation that revolutionizes an industry. Most often, advances in the application of technology in business, science, and other areas of endeavor are a result of the union of many technologies in a way that allows those technologies to leverage one another's benefits. The direction of software component development and use will in large measure be determined by the contributions this area of work makes to the convergence of multiple technologies.

**Software components will spread throughout the computing architecture.**

The greatest benefits from software components will come as components integrate horizontally—meaning with multiple other technologies—and vertically, meaning at multiple levels of the computing model. This chapter looks at some important trends in software components from this dual perspective. It briefly explores how components are integrating with other technologies (such as advanced programming environments, Web servers and browsers, and database management systems) and at multiple levels of the computing model, using as a guide the basic client/server model consisting of a network, a client, and a server.

# Components and the Network

*Components are enabling applications to be spread across the network.*

An often-heard phrase (coined by Sun Microsystems) in the technology industry is that "the network is the computer." This is generally intended to indicate that a network of computers can now be used effectively as a single computing resource. Advances in networking technology now allow computers to share disk drives, memory, and other computing resources.

Based on this infrastructure of shared computing resources, the component approach to software development discussed in this book is allowing applications to be partitioned (divided into pieces), and distributed across networks even though they appear to the user as if they exist solely on the user's individual computer. The following illustration shows a distributed application. In the illustration, components 1 and 2, residing on two separate computers, function together as an application to access the database on another computer. The application is presented with a single interface on the user's computer.

*A distributed application.*

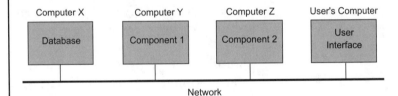

The Internet is by far the most significant factor affecting how these distributed applications built with software components will evolve. The component approach to software development that began on individual computers as an attempt to manage application development through reusable components has now spread to the network to create distributed applications. As "the network" increasingly comes to mean "the Internet," software component technology is converging with Internet technology and is providing exciting new opportunities for globally distributed software applications.

*Components on the Internet enable global delivery of applications.*

At the 1997 Netscape developers' conference, Mark Andreessen, Chief Technology Officer of Netscape, expressed his view of the importance of the convergence of software component technology with the Internet. Andreessen noted that the world is moving toward an "Object Web" in which developers can select from an assortment of components located throughout the Internet to create sophisticated applications. Recognizing the benefits of this approach, Andreessen stated that in Netscape's view, "A standard component model for the Internet is all about creating Internet applications very rapidly."

*Components on the Internet enable global development of applications.*

*An illustration of the concept of the "Object Web." Internet-based applications will be developed by associating components located throughout the Internet.*

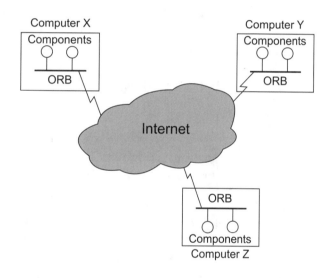

## The Internet: Redefining the Network

Just as the network is rapidly redefining what "the computer" is, the Internet is also redefining what "the network" is. At its inception, the Internet provided a mechanism for researchers to share information across a common computer network. Since then, the Internet has grown dramatically in scope to become a global resource that now includes personal and business uses. Despite this transformation, what has not changed is the Internet's fundamental profile as an "open" network infrastructure.

*The Internet is increasing the scope of "the network."*

The "open" nature of the Internet refers to the use of standardized methods of communication among computers. This use of common standards means that a computer can be connected to the Internet with little effort, and the Internet can then be used as a means to communicate with other computers on the network. This communication can occur despite differences in hardware, software, or other computing factors that have traditionally been impediments to communications among computers.

The Internet uses standard protocols for open access and communication.

Open access has been excellent news for many organizations. Prior to the Internet, for most organizations, "the network is the computer" could only refer to the organization's own network. With the proliferation of network and computing "standards" that were actually proprietary in nature, connection of new computers to the organization's own network was generally difficult, at best; connection of the organization's network with outside networks was nearly impossible. With the advent of the global computing network called the Internet, organizational networks can now be connected to one another and communicate with relative ease.

Open access allows organizational networks to include the Internet.

As organizations expand the scope of "the network" through the use of the Internet, it is natural that the scope of software applications expand accordingly. The sharing of computing resources has grown beyond the organizational network and into the global network of the Internet.

An expanding network allows applications to expand via components.

This means that the component approach to software design and development can be extended to the Internet. The "pieces" of a component-based application can now be distributed on computers across the globe and accessed by users across the Internet. In a very real sense, global applications are now made possible by the convergence of software component technologies and Internet technologies.

## The Internet: Redefining the Business

Business forces are driving software component and Internet technology expansion.

It is important to understand that the convergence of components and the Internet is not simply a technological issue. The significance of this trend is that it is in direct response to the needs of today's organizational and business environment. The expansion of the boundaries of computers, networks, and applications has been driven by the need of organizations to expand their organizational boundaries.

Organizations are extending their organizational boundaries.

In his book *Strategic Information Systems* (Pitman Publishing, 1993), David Targett of the University of Bath, England, refers to the Index Group's attempt to evaluate the organizational impact of technology. In this evaluation, the Index Group refers to "boundary extension" as one of the most significant impacts of the current era of technology. Boundary extension, represented in the following illustration, refers to the expansion of an organization's boundaries to include entities and activities traditionally seen as external to the organization.

*Organizational boundaries are expanding to include entities traditionally seen as external to the organization.*

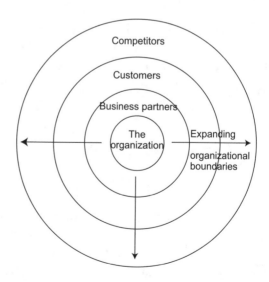

The boundaries of many organizations are expanding to include customers, business partners, constituents, and even competitors. For example, major car manufacturers are implementing computer systems that provide dealers with more control over distribution and delivery. Previous to these types of systems, distribution and delivery was primarily seen as an internal organizational process controlled by the manufacturer. Realizing the benefit of expanding the boundaries of the business and its processes to include dealers, manufacturers are demanding more such systems to support more boundary expansion.

*Organizational boundaries now include customers and competitors.*

Demand on the part of organizations to expand boundaries by extending their information systems preceded the widespread use of the Internet and component-based applications. Most recently, with the existence of the Internet and component-based applications, this boundary expansion is accelerating and taking on new forms.

*Components and the Internet accelerate boundary extension.*

For instance, car dealers and manufacturers are experimenting with extending the previously described distribution and delivery systems beyond the dealer-manufacturer relationship to include customers via the Internet. This would allow customers to select make, model, and color via their home computer. The result is an expansion of the boundaries of the business and its systems into the individual customer's home, and an expansion of customer influence into the previously "internal" systems of the business.

*Organizations are using the Internet to open internal systems.*

## What Does Expansion Mean for GIS?

These trends—the expansion of organizational boundaries, the expansion of the network to include the Internet, and the expansion of software applications to extend across the Internet—have significance for the GIS industry and those who use its technology. By their very nature, software applications that enable the expansion of an organization and its business processes across the Internet involve the distribution, collection, and analysis of information.

*Where information goes, geography goes, and GIS must go.*

Recall from the discussion in Chapter 2 that organizations are now realizing that there is a geographic element in most of this information. Simply put, where the information goes, geography goes; and where geography goes, GIS must go.

Just as the importance of the geographic element of information is driving GIS into the technology infrastructure of many organizations, the extension of this information and its geographic element across the Internet is driving GIS into the realm of globally distributed applications. The software component ap-proach that enables GIS to integrate with the technology infrastructure also provides the foundation for integrating GIS with the Internet.

*Software components are ideal for the Internet.*

The unique nature of the Internet as an "open" network infrastructure is based on its implementation of standard protocols. For software developers to efficiently develop component-based applications, application development frameworks supporting the use of these protocols by components must be created. Many vendors are now creating such frameworks to enable the development of Internet-based distributed component applications.

*GIS components will participate in Internet application frameworks.*

As the demand for GIS to operate on the Internet increases, GIS software components will be required to integrate with these emerging application frameworks. The Netscape Open Network Environment (ONE) is a representative example of what these frameworks look like.

Netscape ONE is defined by Netscape as a platform-independent environment in which applications can be developed, delivered, and run—relying entirely on Internet standard languages (such as HTML and Java) and Internet standard protocols (such as HTTP), using services and development tools provided using these same languages and protocols, and deployed over a standard object request broker (i.e., CORBA).

*Netscape ONE is an example of an Internet application framework.*

The intent of Netscape ONE is to allow applications to be developed as *crossware*—a term used by Netscape to empha-

*Netscape ONE implements an Internet-based component infrastructure.*

size the cross-platform (cross-network, cross-hardware, cross-operating system) nature of distributed component applications. The Netscape ONE model implements, using Internet protocols and standards, all of the elements of a component infrastructure discussed in previous chapters.

These elements include an object request broker (CORBA), an inter-ORB protocol (CORBA IIOP), standard component services providing generic functions such as licensing and security, and support for visual development environments (such as Visual C++) in which components can be developed. Additionally, Netscape ONE emphasizes the use of Internet-focused, cross-platform languages such as Java and JavaScript in the development of components.

*Frameworks will allow GIS to "plug-and-play" on the Internet.*

Software components are evolving to operate within such application frameworks, and GIS software components will evolve to offer the same support. Software developers using components to create applications intended for distribution across the Internet will be able to include GIS components in Internet-focused application development frameworks.

*Software components and the Internet will extend the reach of GIS.*

In an earlier chapter it was stated that "the user base of GIS is becoming much more inclusive. To succeed, GIS must become accessible to a nearly universal user base. Geographic technologies must be usable by anyone working in a digital environment, regardless of their level of understanding of geography or the peculiarities of GIS technology." This goal will be realized as software developers use Internet-based component frameworks to create and assemble components into globally distributed applications that include GIS capabilities.

# Components and the Server

"Server" is defined here as the computer on which either application or data access operations are normally performed, typically in a location geographically distant from the user of the application. This is in contrast to the "client," a computer that is generally focused on the presentation of the application to the user. This is usually the computer with which the user directly interacts. The database access and management aspect of the server is an example of one of the many aspects of the server that are of increasing importance to GIS and GIS software components.

*Trends in server-based data management are important to GIS.*

## Integrating Spatial and Aspatial Data in the Database Server

In most organizations, the database server is typically a DBMS, such as Oracle, Sybase, or Access. In GIS systems, it may be a data management system specifically tailored to spatial information, such as ESRI's Spatial Database Engine. In GIS systems that require integration with existing repositories of information stored in a DBMS, the spatial data management system and the DBMS may work in tandem. A significant trend, however, is the integration of the two data management systems into a single system that manages both spatial and aspatial data.

*Spatial and aspatial data are merging in the DBMS.*

The integration of spatial and aspatial data into a single repository parallels the integration of GIS software technology via software components. Organizations recognizing the increasing importance of geographic information have initially demanded the integration of geographic analysis into critical software applications. Geographic components are developing to satisfy that need.

*Data-level integration of GIS parallels application-level integration.*

As the usefulness of geographic information is being proven by this integration at the application level, organizations are demanding integration at the data level to make the geographic data itself as accessible, manageable, and secure as any other important information asset. This integration is represented in the following illustration.

*Integrating spatial data with corporate data repositories parallels the move toward integrating GIS software with corporate information systems.*

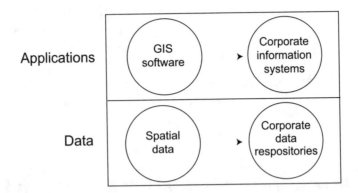

Add the increasing size of geographic databases to these demands for accessibility, manageability, and security, and the result is a movement toward the integration of spatial data with aspatial data using a common set of data management techniques and tools. Because of the DBMS industry's many years of experience in managing very large databases to ensure that they are accessible, well-managed, and securely protected, this trend is moving geographic data closer to, and in some cases into, the DBMS itself.

Organizations see DBMSs as having proven experience.

## Components and the Database Server

Data management can be encapsulated in components.

As discussed in previous chapters, the concept of software components is based on the premise that specific functions that might be useful in an application can be identified with specific objects, and these objects can be collected into components. This isolation of functionality into logical, manageable units can be applied to the database server as well. The functions required to access and manage the database can in many cases be encapsulated into components.

Recognizing this potential to apply software component concepts to DBMS architectures, the leading DBMS vendors are providing data management capabilities using component-like models. This approach relieves software applications of the responsibility of data-specific tasks by providing the applications with components that specialize in accessing the database server. In addition, custom data management components can be developed by application developers and integrated with the DBMS vendor's component-compliant architecture.

*DMBS vendors are implementing component frameworks.*

Oracle's Network Computing Architecture (NCA) provides an example of such an approach. In a clear statement of how NCA exemplifies the database vendors' recognition of the importance of the component approach, Oracle states that the "Network Computing Architecture provides a clear path to integrating client/server computing with the Internet and distributed object architectures." ("Network Computing Architecture," Oracle White Paper, September 1996).

*Oracle's NCA is an example of a DBMS component framework.*

The NCA is built on the concept of components, using a CORBA-compliant object request broker. Software components (called "cartridges" by Oracle) can access the services of the Oracle DBMS by using the object request broker to interact with database components. The following illustration shows a typical example of DBMS and component model integration.

*An example of DBMS integration with a component model.*

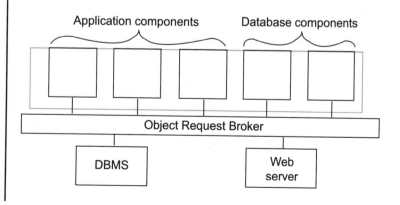

The previous illustration was adapted from Oracle's NCA diagrams to show how products such as Oracle NCA integrate the DBMS within a component model. Application components running on a client or a server operate through an object request broker (called the "inter-cartridge exchange" in NCA terminology) to communicate with database components (called "data cartridges" in NCA) that supply database services. Database components also use the object request broker to communicate with the DBMS (the "universal server" in NCA), to satisfy the data requests of the application components. Clients and application components may also communicate, via the object request broker, with the Web server, which can act as an Internet application server (called the "universal application server" in NCA).

*NCA integrates the DBMS with an object request broker.*

Oracle's efforts represent a general trend toward enabling DBMSs for the component world. As DBMS vendors implement their databases within component frameworks such as Oracle's NCA, the database will become increasingly accessible to component-based applications. Using a component approach, DBMSs will also become extendable. For example, in Oracle's NCA, new components can be created by application developers and added to the NCA to create support in the database for new data types, including spatial data.

*Components enable DBMSs to be more "open" and "extendable."*

## What Do DBMS Component Models Mean for GIS?

For the GIS industry and those relying on its technologies, the most obvious significance of this trend is the movement of spatial data management away from mechanisms proprietary to the GIS industry. DBMS vendors are working to support many data types not previously supported by typical relational database management systems, such as video, audio, and text. These DBMS vendors see spatial data as simply another data type. As a result, DBMS ven-

Spatial data management will move toward DBMSs.

dors are beginning to include options for storing and managing spatial data as another data type supported by the DBMS.

The effect of this trend is that the management of spatial data will have heavier involvement from DBMS vendors and may perhaps even transition away from GIS vendors to DBMS vendors. To the degree that this occurs, vendors of GIS technology will focus on tools enabling the analysis and presentation of geographic information. As database-level issues are increasingly addressed by DBMS vendors, GIS vendors can focus on application-level issues such as the development of components that integrate GIS technology at the application level.

GIS vendors will focus on data analysis and presentation.

Components specialized for spatial data access and management will provide software developers with "plug-and-play" access to spatial data stored in the DBMS environment. Just as GIS software components are making GIS functions accessible to traditional software applications, component-based access to spatial data within a DBMS will make spatial data available to a wider set of users and applications.

Components will provide "plug-in" access to spatial data in DBMSs.

In a manner similar to the way in which SQL and ODBC freed application developers from concerning themselves with proprietary data storage mechanisms in particular DBMSs, components specialized for spatial data access will free developers from concern with proprietary mechanisms for storing spatial data. These benefits will accrue as DBMS technologies and spatial data management technologies converge and integrate with the component model.

# Components and the Client

As mentioned in the discussion of the server, the term *client* generally refers to that which performs presentation-focused operations, and is usually the computer with which the user directly interacts. It is frequently the access

New types of devices are beginning to act as clients.

point into the system for the user. Perhaps the most significant trend affecting the use of components as they relate to the client is the redefinition of the term *client* to include devices beyond the typical computer. One example of these new types of clients is the information appliance.

# Information Appliances

Information appliances integrate technologies for consumers.

The concept of information appliances may be defined as the integration of multiple technologies to create commonly used consumer items. Internet-aware telephones and digital cameras are two examples of information appliances that have emerged as technologies have converged.

An example of how geographic-related technologies are already being integrated with such information appliances is the inclusion of GPS technology in digital cameras. When a photograph is taken, the GPS determines the geographic location at which the photograph was taken, and stores it. The location can then be downloaded with the photograph. The following illustration depicts the convergence of multiple technologies in an appliance.

*An example of multiple technologies converging in an appliance to create an "information appliance."*

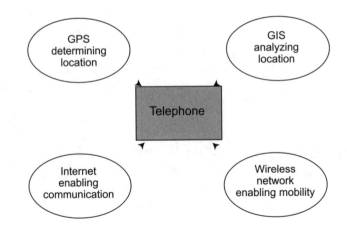

Of interest here is the role that Web servers are beginning to play in the development of information appliances. A Web server receives and processes client requests sent via the Internet. It functions as the contact point for an Internet browser issuing a request. The most typical task of a Web server is to respond to requests for an HTML document by locating the document and sending it to the client requesting it.

**Web servers process requests on the Internet.**

In recent years, Web servers have evolved to include capabilities for passing requests and responses between components by using object request brokers. This capability of facilitating communication between small software components, combined with miniaturization of the Web server, has led to exploration of the Web server as a control mechanism for a wide variety of household and industrial products. In these cases, the Web server itself can be implemented as a software component within a component framework.

**Web servers now work with component ORBs.**

For example, a miniaturized Web server might be embedded in a residential gas meter. This would allow a gas provider to monitor the meter remotely via the Internet rather than sending a meter reader to a residence. Weiser Lock is experimenting with Web servers embedded in door locks to allow remote changing of lock codes, and Xerox plans to embed a Web server into its copiers to allow status and job monitoring. The small size of the software in these appliances and their ability to interact with other such appliances enables them to act as clients as well as servers, receiving and sending requests as necessary.

**Miniaturized Web servers are being embedded in traditional appliances.**

The result of such trends is that the definition of what can be a "client" or a "server" is changing. A device no longer needs to be a computer in the traditional sense to function as a client or server, it simply needs the minimum amount of intelligence and processing power required to communicate with another client or server and to accomplish a specific action. As devices such as information appliances

*Client* and *server* are both increasingly inclusive terms.

emerge, software is being required to run in increasingly constrained environments with very limited amounts of computing resource.

The limited computing resources available in devices such as information appliances means that software developed to support them must use the extreme minimum of resources, supply only the most basic functions required to perform necessary operations, and be completely transparent once embedded in the appliance. Because these are exactly the demands that today's software components are intended to satisfy, software components are poised to be a significant contributor to the development of these information appliances.

Software components add the intelligence to information appliances.

## What Does Appliance Technology Mean to GIS?

GIS technology can provide significant advantages in the emerging world of information appliances. For instance, you might determine the location of a portable phone using an embedded GPS system and pass this information to the person being called to draw a map of the caller's location. Or perhaps the caller's own location would be drawn on the phone's own miniature display to show a major cross-street map for navigating the immediate vicinity.

GIS on a portable phone?

Such uses of GIS technology may be barely believable to those who are accustomed to traditional GIS systems of the earlier eras of GIS technology. As discussed in Chapter 2, the emerging infrastructure era of GIS is in response to demands for the application of GIS technology at the most basic levels of activity. This not only includes activities at the level of organizational infrastructure, but at the level of personal activities, as in the portable phone example just described.

GIS will participate in the most basic of business and personal activities.

Only GIS software components meet the demands for such uses of GIS software, and are providing the foundation for these uses. Software components and the infra-structure to support them are evolving to meet demands that range from corporate information systems assembled from many components to individual components operating on information appliances. GIS software, via software components, will participate in this evolution to make useful applications of GIS technology available to an unprecedented number of users.

GIS in slim clients will make GIS nearly universally available.

# Components and You

Diverse roles are affected by component technology.

The advent of software components and the development of GIS software components and applications using them are having significant impacts on individuals at many levels. In both personal and professional life, software components and the advances surrounding them are increasing both the opportunities and the challenges of effectively using information technology.

## What Does All of This Mean for You?

Business managers will find increased flexibility within component technology applicable to functions and processes.

*Business managers* will find that software components and their convergence with other technologies provide increased flexibility for the business and its processes. Rapid changes to business processes may become less difficult when systems supporting those processes are built using components that can be easily replaced by new components that support the business change.

This means that the organization, by way of its information systems, will have the ability to quickly provide new products and services, to rapidly respond to new business opportunities, and to do both with a higher level of quality. Changes in business scope—the boundary extension referred to earlier in this chapter—can be enabled by the convergence of software components with technologies such as the Internet.

Business managers will be challenged to be technology aware.

There are many challenges for business managers in regard to technology, but perhaps the most important is that of *awareness*. They must stay in touch with new technologies and how these technologies can benefit their organizations. The advent of components combined with the increasing relevance of geographic information will provide new strategic opportunities.

Business managers must remain aware of these technology trends and consider how the capabilities provided by them can enhance business strategy. Those who do so will discover new business opportunities and seize those opportunities more quickly than managers who are not aware of these important trends.

*Technical managers* will find that software components add a new element to the technical portfolios with which they can respond to requests for information systems. Software component benefits such as maximizing the portability of applications to multiple computing platforms, reduction of application development backlogs, and higher levels of reusability provide technical managers with new opportunities when designing the information infrastructure and technical strategies of their organizations.

Technical managers will find new capabilities.

Technical managers designing technical strategies and infrastructures with components in mind will position themselves for future technologies based on component models. The case of Purolator Courier Ltd. cited in Chapter 1 illustrates how the concept of components is enabling organizations to implement new technical architectures.

The benefits of software component technology, however, come with another challenge. Software components will increasingly enable application development to occur further down the technology supply chain. This suggests that in some cases members of a technical organization's user base will become component assemblers developing

Technical managers will be challenged to redefine developer/user relationships.

Technical managers will be challenged to explore the application of GIS.

GIS professionals will make GIS more widely available.

GIS managers and professionals will face the challenge of learning non-GIS technologies.

their own solutions. This blurs the line between technology groups and their clients, and forces technical managers to redefine their relationship with their clients.

As with business managers, technical managers must see geographic technologies as offering new possibilities for providing solutions. The geographic element provides an entirely new dimension of information to work with, and geographic technologies such as GIS software components are making this dimension much more accessible. Because of this, technical managers must be mindful of the contribution geography can make when they are discussing strategic issues with business and corporate managers, or when they are discussing development requests with users.

*GIS managers and GIS professionals* will find that software components provide new opportunities for integrating GIS with other information systems. This will amount to the "mainstreaming" of GIS for many organizations. In addition, when used to develop even the more traditional, standalone GIS implementations, a component approach can deliver GIS implementations that use industry-standard user interfaces and therefore look and feel like other computing applications with which users are familiar. These capabilities provided by GIS software components will enable GIS managers and professionals to make GIS technologies and their benefits more widely available.

A major challenge facing GIS managers and professionals is the integration of non-GIS technology concepts and skills into their GIS organizations. GIS managers must understand the importance of enabling their groups to acquire new tools and skill sets that may not typically be associated with GIS. These may include visual programming environments such as Visual Basic and Delphi, and Internet technologies and skills such as Java and multimedia design.

Just as technical managers are faced with a dilemma as software components blur the line between technology groups and their clients, some GIS managers will find the distinction between the GIS and IS organizations becoming less clear. IS professionals will begin creating applications once considered within the domain of GIS professionals, and vice versa. GIS professionals will face a learning curve in gaining an understanding of the software engineering concepts and practices that must be learned in order to understand and use GIS software components.

*They will face the challenge of redefining GIS/IS organizational relationships.*

*Software developers* will find new tools for delivering useful applications. Software components will bring an increase in the number of "solution developers," whose job is to bridge the gap between software and business processes by assembling components into business systems. In some cases, these solution developers may even be end users. GIS software components will add yet another set of capabilities—namely, geographic analysis—to the suite of tools with which developers, and now users, can create useful applications.

*Software developers will find new capabilities and solution-oriented tools.*

Some software developers will be faced with the challenge of either ceding some amount of control of application development to solution developers, who may even be typically seen as users, or to join the ranks of solution developers. With GIS software components in particular, software developers and solution developers face the challenge of learning geographic concepts, such as coordinate systems and cartographic projections, in order to effectively work with tools incorporating GIS technology.

*Software developers will be challenged with learning geographic concepts.*

*Users* will find that geographic information and the technology to use it becomes increasingly accessible as GIS software components allow GIS to be integrated with everything from portable phones to ATM machines. Just as spreadsheet software has become a standard tool on nearly every desktop—making accounting and financial

Users will find increased accessibility of geographic information.

analysis accessible to office workers—GIS technologies will make geographic analysis accessible across the enterprise. The ultimate intent of those who work with GIS software components is to reduce the challenges faced by users to only those related to the creative use of geographic information.

# Appendix A

## Other MapObjects Products

ESRI has produced additional products related to the version of MapObjects discussed in this book. These products include the MapObjects Internet Map Server (IMS) for serving maps on the Internet, and a "light" version of MapObjects for lower-cost, lesser-function uses. This appendix briefly introduces these products.

## MapObjects LT

MapObjects LT is the "light" version of MapObjects. The product is intended to service a market for those who need only basic mapping tools rather than the more full-featured MapObjects product. The pricing structure of MapObjects LT reflects this difference as well: the product is currently offered at a lower cost than the full version of MapObjects and with no run-time (royalty) fees.

The differences between MapObjects and MapObjects LT may be viewed from two perspectives. First, there are differences in the functional capabilities of the two products. Second, and underlying these functional differences, there are differences in the object models of the two products. The following diagram is adapted from ESRI's Web site. It shows the functional differences between MapObjects 1.1 (the current version of the full MapObjects product as of this writing) and MapObjects LT.

*A comparison of MapObjects and MapObjects LT (source: ESRI Web site).*

| MapObjects 1.1 | LT | Feature/Function |
|:---:|:---:|---|
| ✔ | ✔ | Pan and zoom through multiple map layers. |
| ✔ | ✔ | Display a wide variety of image formats. |
| ✔ | ✔ | Display and query shapefiles. |
| ✔ | | Edit shapefiles. |
| ✔ | | Create new shapefiles. |
| ✔ | | Display ARC/INFO coverages. |
| ✔ | | Display SDE layers. |
| ✔ | ✔ | Label map features. |
| ✔ | ✔ | Retrieve information about map features. |
| ✔ | ✔ | Create value maps, classification maps, and graduated symbol maps. |
| ✔ | | Create dot density maps. |
| ✔ | ✔ | Select features using an expression. |
| ✔ | | Select features using spatial operators. |
| ✔ | ✔ | Select features using the mouse. |
| ✔ | | Process record sets. |
| ✔ | | Perform address geocoding. |
| ✔ | | Event tracking layer (GPS integration). |
| ✔ | | Support MapObjects Internet Map Server extension. |
| ✔ | ✔ | Includes ESRI's Data & Map CD, a compilation of over 1.2 Gb of geographic data, including states, counties, cities, roads, landmarks, metropolitan statistical areas, ZIP codes, census tracts, countries, and much more. |

The differences between the two packages lie in the elimination from MapObjects LT of three categories of GIS capabilities found in MapObjects: spatial analysis, advanced data access, and spatial data editing.

- ❏ **Spatial Analysis.** MapObjects LT does not contain MapObjects' *address matching* objects. This means that MapObjects LT does not provide functions for easy determination of the location of an address. *Spatial selection* of features is also removed from MapObjects LT. In contrast, the full version of MapObjects includes options such as searching for features that intersect another feature, are contained by another feature, or are within a specified distance of another feature. Also eliminated from MapObjects LT is the ability to perform *real-time tracking* (e.g., integrating GPS) using the TrackingLayer object found in the full version of MapObjects.

- ❏ **Advanced Data Access.** The full version of MapObjects is capable of *accessing ARC/INFO coverages and ESRI's Spatial Database Engine (SDE)*. These capabilities have been removed from the LT version. Also removed from MapObjects LT is the ability to perform *recordset processing*. This means, for example, that neither the tables associated with a shapefile nor their values can be edited.

- ❏ **Spatial Data Editing.** In contrast to MapObjects, MapObjects LT is not capable of *creating new shapefiles*. It also lacks the ability to *edit shapefiles*.

To accomplish these modifications, the object model of MapObjects was modified. Some objects—such as the TrackingLayer, Table, Statistics, and DotDensityRender objects—have been removed completely. Others have been removed, but some of their capabilities have been incorporated into other parts of the object model. If you wish to compare diagrams of the object models of MapObjects and MapObjects LT, you will find a diagram for MapObjects on the enclosed CD-ROM, and a diagram for MapObjects LT on the ESRI Web site (*www.esri.com*).

For instance, in MapObjects the LabelRenderer object draws text next to a feature, with the text value taken from the feature attribute table (see Chapter 8). The LabelRenderer object has been removed completely from MapObjects LT, but some of its properties have been modified and moved to the MapLayer object. For example, the LabelRenderer's Field property, which specifies the field from which to take text for display, has become the LabelField property of the MapLayer object. In this way, a minimal amount of LabelRenderer functionality has been retained in MapObjects LT.

This difference in the object models is important to developers. In some cases, applications developed in MapObjects LT may have new requirements imposed on them over time. If the new requirements demand capabilities found only in the full version of MapObjects, the application's use of MapObjects LT must be replaced with MapObjects.

This may require some rewriting of existing code because of changes in the object model. Recalling the previous LabelRenderer example, an application using the LabelField property of the MapLayer object in MapObjects LT will need to be modified to reference the Field property of the LabelRenderer object in MapObjects.

## MapObjects Internet Map Server

The MapObjects Internet Map Server (IMS) is a product that can be used along with MapObjects in order to deliver interactive maps from an Internet Web site. The product consists of software that extends the Web server to communicate with a MapObjects application, and a software component (called the WebLink control) that enables a MapObjects application to listen for and respond to requests from the Web server extension. Together, these two components function as a two-way bridge between the Internet and a MapObjects application. The following illustration depicts the use of MapObjects IMS.

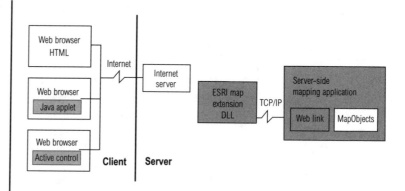

*A diagram of a MapObjects IMS implementation (source: "Getting Started with MapObjects IMS," ESRI).*

The MapObjects IMS Web server extension is implemented as a DLL in two versions: NSAPI and ISAPI. The NSAPI (Netscape Server Application Programming Interface) DLL supports Netscape Server or other Web servers that support NSAPI. The ISAPI (Internet Server Application Programming Interface) DLL supports Microsoft Internet Information Server (IIS) or other Web servers that support ISAPI, such as Microsoft Personal Web Server (PWS) for Windows 95. The MapObjects IMS WebLink control is implemented as an ActiveX control that can be added as a component into any application development environment that supports ActiveXs.

To access an application using MapObjects IMS, the Web browser first issues a request to the Web server by sending it a URL. The URL must include specific parameters indicating that it is a request to be sent to the IMS server extension. The URL also packages information about the map being requested by the client. This information is later used by MapObjects to create the map. The following is a sample of a URL sent from a Web browser to a Web server to request a map from a MapObjects application using IMS:

```
http://www.ruis.org/scripts/.esrimap?name=ruismap&cmd=map
```

The first portion of the URL, http://www.ruis.org, directs the request toward a specific Web server on the Internet. The second portion, /scripts/.esrimap?name=ruismap, tells the Web server that this is a request to be passed to the IMS Web server extension (named esrimap) and supplies the name (ruismap) of the MapObjects application with which the IMS server extension is to communicate.

The remaining portion of the URL, &cmd=map, supplies an argument (cmd) and a value (map) that will be used by the MapObjects application in creating a map. Any number of argument/value pairs may be embedded in the URL to further define the map request.

After receiving the URL from the browser, the Web server passes the request on to the IMS Web server extension, and the extension establishes communication with the indicated MapObjects application, which in the foregoing URL example is ruismap. The extension then passes the map arguments and values contained in the URL to the IMS WebLink control embedded in the specified MapObjects application. These arguments and values are received by the WebLink control in an event named "Request," which is fired when the Web server extension contacts the control.

The programmer creating the mapping application can insert code into the Request event procedure to process the argument/value pairs, which may include calling other procedures for additional processing. When the MapObjects application has finished creating the map, the map is then converted to an image in one of several standard formats.

The programmer then uses methods of the WebLink object, such as the WriteString method, to send the image and any accompanying HTML back to the IMS Web server extension. The IMS Web server extension then sends it to the Web server, which sends it back to the client browser for display.

*A sample of an Internet mapping application using MapObjects Internet Map Server.*

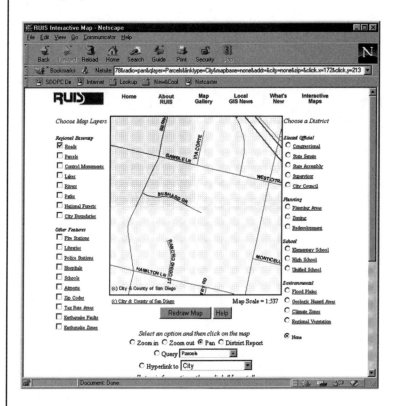

The map itself can be made interactive by indicating in the accompanying HTML that the image is clickable. Interactivity can be enhanced through the use of Java or client-side ActiveX controls. ESRI supplies Java code and ActiveX controls to assist programmers who wish to develop maps with these additional capabilities.

# Appendix B

## For Further Reference

This appendix contains references to Web site, book, and periodical resources. Web site resources include those for general software components, GIS-specific software components, and MapObjects Internet Map Server.

## Web Sites

The following are Web sites you may find helpful in exploring software components further. They are listed alphabetically by Web site address.

## General Software Components

**http://www.microsoft.com**

The Web site of Microsoft, creator of the Component Object Model (COM), Distributed COM (DCOM), OLE, and ActiveX. It may be a challenge finding focused information on these topics because the site is huge and addresses many other issues. URLs change regularly on this site; therefore, start by searching for some of the articles in the last section of this appendix.

http://www.netscape.com

The Web site of Netscape, creators of the Navigator Web browser and Netscape Web servers. Look for information on Netscape Open Network Environment (ONE) as an example of a component framework built using Internet standards and centered around an ORB model.

http://www.omg.org

The Web site of the Object Management Group (OMG), creators of the Common Object Request Broker Architecture (CORBA). A wealth of information on CORBA for beginner and expert alike.

http://www.opengis.org

The Web site of the Open GIS Consortium (OGC), an industry group focusing on standardization and interoperability in the GIS industry. Contains the publications of the OGC and other helpful information on OGC activities.

http://www.oracle.com

The Web site of Oracle Corporation, makers of the Oracle DBMS. Look for information on their Network Computing Architecture (NCA) as an example of a database-oriented component framework built around a component object request broker (ORB). Includes articles listed later in this appendix.

## GIS Software Components

The following are Web sites of several vendors who have created component-based GIS software products. There are many vendors now producing such products, with more joining the ranks each day. This list is intended to be representative rather than comprehensive.

**http://www.bluemarblegeo.com**

The Web site of Blue Marble Geographics, Inc., a company that takes a component approach to all of its products. In particular, search for information on the *GeoView* and *GeoViewLT* products.

**http://www.esri.com**

The Web site of Environmental Systems Research Institute, makers of the *MapObjects, MapObjectsLT*, and *Internet Map Server* (IMS) products.

**http://www.intergraph.com**

The Web site of Intergraph. Search for information on the *GeoMedia* product. Also search for information on *Jupiter*, a project launched by Intergraph to develop a component approach to their software products. GeoMedia is a result of the Jupiter project.

**http://www.mapinfo.com**

The Web site of MapInfo Corporation, one of the first vendors to support GIS on desktop PCs. Search for information on the *MapX* component-based mapping product.

**http://www.sylvanmaps.com**

The Web site of Sylvan Ascent, Inc., makers of *Sylvan Maps*, one of the very first GIS software components.

**http://www.visualcomp.com**

The Web site of Visual Components, Inc., a subsidiary of Sybase. Sybase is the maker of the Sybase DBMS and owner of Powersoft, makers of the PowerBuilder application development tool. Search for information on *GeoPoint*, one of the first GIS software components from a mainstream (i.e., non-GIS) computing vendor.

## Web Sites Using the MapObjects Internet Map Server

Because this book included a brief overview of the MapObjects Internet Map Server (IMS) to describe how ESRI, like many other vendors, has extended its GIS software component to the Internet, here are a few sites demonstrating IMS.

**http://maps.esri.com**

A Web site containing several examples using IMS and developed by ESRI. Includes some samples using Java.

**http://www.esri.com/base/products/internetmaps/ visit_sites.html**

A Web site with links to sites of ESRI clients who have developed IMS applications.

**http://gis.ci.ontario.ca.us/gis/index.htm**

A winner in ESRI's 1997 User Conference competition for Internet mapping using IMS.

**http://www.ruis.org**

Another winner in ESRI's 1997 User Conference competition for Internet mapping using IMS.

# *Books*

## Software Component Issues

This list provides a few select titles helpful to those interested in further reading on the software component topics discussed in this book, particularly the various component architecture alternatives such as CORBA and DCOM.

David Chappel, *Understanding ActiveX and OLE: A Guide for Developers and Managers* (Microsoft Press, 1996).

An introduction to OLE and ActiveX. This is a good conceptual introduction to the significance of COM and related technologies as part of Microsoft's strategy for components. It is at a technical level appropriate for technically oriented decision makers.

Robert Orfali, Dan Harkey, and Jeri Edwards, *The Essential Distributed Objects Survival Guide* (John Wiley & Sons, 1996).

A comprehensive tour of distributed objects and components. Excellent technical explanations (in plain language) of various component models with an emphasis on CORBA. Includes a comparison (admittedly biased) of CORBA and COM.

Robert Orfali, Dan Harkey, and Jeri Edwards, *Instant CORBA* (John Wiley & Sons, 1997).

There is a lot of overlap between this book and *The Essential Distributed Objects Survival Guide*, written by the same authors, but enough new material to make it worthwhile reading. It contains updates on CORBA 2.0, future developments in CORBA, and discussion of how CORBA/IIOP, Java, and the Internet are coming together to support the "object Web."

Dale Rogerson, *Inside COM* (Microsoft Press, 1997).

Destined to become a COM classic, this is a technical overview of COM presented in an understandable way.

David Taylor, *Object-Oriented Technology: A Manager's Guide* (Addison-Wesley, 1990).

The ultimate manager's guide to object technology. A high-level, understandable explanation of what objects are and how they are used. This is an excellent introductory book to object basics.

## GIS Software Components

The OGIS Project Technical Committee of the Open GIS Consortium, Inc., Kurt Buehler and Lance McKee, editors, *The OpenGIS Guide: Introduction to Interoperable Geoprocessing* (Open GIS Consortium, Inc., 1996).

Available at the OGC's Web site (*http://www.opengis.org*). An excellent technical guide to the OGIS specification and a "must read" for those interested in distributed, interoperable GIS. The OGIS specification itself is also available at the OGC's Web site.

## *Articles*

There is a wealth of articles on software components from various sources and perspectives. The following are a few select articles that provide further information on specific topics discussed in this book.

## Software Component Issues

Brockschmidt, Kraig, "What OLE Is Really About" (Microsoft, July 1996).

A technical overview of the problems OLE and COM are intended to solve and how they solve them. Available at Microsoft's Web site.

Kindel, Charlie, "Industrial Strength OLE: Using Microsoft's Object Technology to Build Business Solutions" (Microsoft Developer Network News, 1996).

A very brief introduction to how Microsoft is assisting in the development of industry-specific component standards.

Microsoft, "The Microsoft Object Technology Strategy: Component Software" (Microsoft, 1997).

Why Microsoft thinks components are important. Available at Microsoft's Web site.

Pope, Alan L., "CORBA" (Object Management Group). A more in-depth overview than Schmidt's article.

Well formatted and an easy, though technical, read. Available on OMG's Web site in the "CORBA for Beginners" section.

Schmidt, Doug, "Overview of CORBA" (Object Management Group, July 1997).

A good overview of the major components of CORBA, with definitions of each. Available on the OMG's Web site in the "CORBA for Beginners" section.

Udell, Jon, "Componentware" (*Byte Magazine*, May 1994).

Written at a time when Microsoft was transitioning from VBXs to OCXs, this article provides a brief overview of the work of various vendors laying the foundations at that time for software components.

## GIS Software Components

As yet there are not many articles on the subject of GIS software components, but the following are several that are helpful.

Environmental Systems Research Institute (ESRI), "The Future of GIS on the Internet: An ESRI White Paper" (ESRI, 1997).

A conceptual discussion of a possible architecture for the application of GIS software components on the Internet. Available at ESRI's Web site.

Ganter, John, "Arcview and MapObjects: An Architectural Comparison for Developers" (Sandia National Laboratories).

A comparison that touches on the difference between a component approach and a traditional GIS toolkit approach. Available at Sandia National Laboratories' Web site.

GIS World, "The Open GIS Connection," (*GIS World,* 1995-).

A column written by representatives of the OGC discussing the OGC's work on interoperable GIS. The excellent articles from this column are available at the OGC's Web site (*http://www.opengis.org*).

# Index

# More OnWord Press Titles

## Computing/Business

Lotus Notes for Web Workgroups
$34.95

Mapping with Microsoft Office
$29.95 Includes Disk

The Tightwad's Guide to Free Email
and Other Cool Internet Stuff
$19.95

## Geographic Information Systems (GIS)

GIS: A Visual Approach
$39.95

The GIS Book, 4E
$39.95

GIS Online: Information Retrieval, Mapping,
and the Internet
$49.95

INSIDE MapInfo Professional
$49.95 Includes CD-ROM

Minding Your Business with MapInfo
$49.95

MapBasic Developer's Guide
$49.95 Includes Disk

Raster Imagery in Geographic Information
Systems Includes color inserts
$59.95

INSIDE ArcView GIS, 2E
$44.95 Includes CD-ROM

ArcView GIS Exercise Book, 2E
$49.95 Includes CD-ROM

ArcView GIS/Avenue Developer's Guide, 2E
$49.95  Includes Disk

ArcView GIS/Avenue Programmer's
Reference, 2E
$49.95

ArcView GIS /Avenue Scripts: The Disk, 2E
Disk $99.00

ARC/INFO Quick Reference
$24.95

INSIDE ARC/INFO, Revised Edition
$59.95 Includes CD-ROM

Exploring Spatial Analysis in Geographic
Information Systems
$49.95

Processing Digital Images in GIS:
A Tutorial for ArcView and ARC/INFO
$49.95

Cartographic Design Using ArcView GIS and
ARC/INFO: Making Better Maps
$49.95

# Softdesk

*INSIDE Softdesk Architectural*
$49.95 Includes Disk

*INSIDE Softdesk Civil*
$49.95  Includes Disk

*Softdesk Architecture 1 Certified Courseware*
$34.95  Includes CD-ROM

*Softdesk Civil 1 Certified Courseware*
$34.95 Includes CD-ROM

*Softdesk Architecture 2 Certified Courseware*
$34.95  Includes CD-ROM

*Softdesk Civil 2 Certified Courseware*
$34.95 Includes CD-ROM

# MicroStation

*INSIDE MicroStation 95, 4E*
$39.95 Includes Disk

*MicroStation for AutoCAD Users, 2E*
$34.95

*MicroStation 95 Exercise Book*
$39.95 Includes Disk
Optional Instructor's Guide $14.95

*MicroStation Exercise Book 5.X*
$34.95 Includes Disk
Optional Instructor's Guide $14.95

*MicroStation 95 Quick Reference*
$24.95

*MicroStation Reference Guide 5.X*
$18.95

*MicroStation 95 Productivity Book*
$49.95

*MicroStation for Civil Engineers:*
*A Design Cookbook*
$49.95

*Adventures in MicroStation 3D*
$49.95  Includes CD-ROM

*101 MDL Commands (5.X and 95)*
Executable Disk $101.00
Source Disks (6) $259.95

# Pro/ENGINEER and Pro/JR.

*Automating Design in Pro/ENGINEER*
*with Pro/PROGRAM*
$59.95 Includes CD-ROM

*Pro/ENGINEER Tips and Techniques*
$59.95

*INSIDE Pro/ENGINEER, 3E*
$49.95  Includes Disk

*INSIDE Pro/JR.*
$49.95

*Pro/ENGINEER Exercise Book, 2E*
$39.95  Includes Disk

*INSIDE Pro/SURFACE: Moving from Solid*
*Modeling to Surface Design*
$90.00

*Pro/ENGINEER Quick Reference, 2E*
$24.95

*FEA Made Easy with Pro/MECHANICA*
$90.00

*Thinking Pro/ENGINEER*
$49.95

# Other CAD

*Fallingwater in 3D Studio*
$39.95  Includes Disk

*INSIDE TriSpectives Technical*
$49.95

# SunSoft Solaris

*SunSoft Solaris 2.\* for Managers and
Administrators*
$34.95

*SunSoft Solaris 2.\* User's Guide*
$29.95  Includes Disk

*SunSoft Solaris 2.\* Quick Reference*
$18.95

*Five Steps to SunSoft Solaris 2.\**
$24.95  Includes Disk

*SunSoft Solaris 2.\* for Windows Users*
$24.95

# Windows NT

*Windows NT for the Technical Professional*
$39.95

# HP-UX

*HP-UX User's Guide*
$29.95

*Five Steps to HP-UX*
$24.95  Includes Disk

# OnWord Press Distribution

## End Users/User Groups/Corporate Sales

OnWord Press books are available worldwide to end users, user groups, and corporate accounts from local booksellers or from SoftStore Inc. Call toll-free 1-888-SoftStore (1-888-763-8786) or 505-474-5120; fax 505-474-5020; write to SoftStore, Inc., 2530 Camino Entrada, Santa Fe, New Mexico 87505-4835, USA, or e-mail orders@hmp.com. SoftStore, Inc., is a High Mountain Press company.

## Wholesale, Including Overseas Distribution

High Mountain Press distributes OnWord Press books internationally. For terms call 1-800-4-ONWORD (1-800-466-9673) or 505-474-5130; fax to 505-474-5030; e-mail to orders@hmp.com; or write to High Mountain Press, 2530 Camino Entrada, Santa Fe, NM 87505-4835, USA.

## Comments and Corrections

Your comments can help us make better products. If you find an error, or have a comment or a query for the authors, please write to us at the address below or call us at 1-800-223-6397.

## OnWord Press, 2530 Camino Entrada, Santa Fe, NM 87505-4835 USA

## On the Internet: http://www.hmp.com